A Bushman Remembers

*For my parents,
Jim and Ivy Mahoney*

"Tarboy" – *James Arthur Mahoney*

A Bushman Remembers
Life on the land when horsepower held sway

by

Tarboy
(James Arthur Mahoney)

Edited by James S Mahoney

Central Queensland
UNIVERSITY
PRESS

© James S Mahoney, 2003

This book is copyright. Apart from any fair dealing for the purpose of private study, research, criticism or review, as permitted under the Copyright Act of 1968, no part may be reproduced by any process without written permission. Enquiries should be made to the publisher.

First published in 2003 by:
Central Queensland University Press (Outback Books)
PO Box 1615,
Rockhampton
Queensland 4700
Phone: 07 4923 2520
Fax: 07 4923 2525
Email: d.myers@cqu.edu.au
Website: www.outbackbooks.com

National Library of Australia
Cataloguing-in-Publication Data:

James Arthur Mahoney (Tarboy).
A bushman remembers : Life on the land when horsepower held sway.

ISBN 1 876780 25 8.

1. Country life - Australia - Anecdotes. 2. Farm life - Australia - Anecdotes. I. Mahoney, James S. II. Title. III. Title : Land (Sydney, N.S.W.).

630.994

Typeset in 12 on 14 point Monotype Columbus by Frank Povah at The Busy Boordy, Dunalley

Printed by CQU Publications Unit, Rockhampton, Qld

Cover painting by Edith Neish

Cover photograph THE LAND HERITAGE COLLECTION

The author and publisher wish to thank Rural Press Pty Ltd for that company's kind permission to reproduce photographs from THE LAND HERITAGE COLLECTION

FOREWORD

There is a rich common thread running through the writings of James Arthur Mahoney in the column "Tarboy" which provides an educative essence in relation to our great bush heritage. You can sense the Mahoney frustration that in the bout of modernisation which took place in Australia after World War II, a good deal of folklore and bush heritage was being lost to the new generations.

This was certainly true for the 'bash through' generation of the Baby boomer Beatles period when the bulldozers dominated in more ways than one. Mahoney effectively challenged this and subsequent generations to know more about aspects of the past. It is a bonus that Mahoney in fact jumps further back before his own formative years to pick up on the first Sydney Show at Parramatta around 1820, then through to many of his own experiences such as the first day at school and onto more modern things.

There is the Outback and the near Outback and, if you like, the Inback areas with close proximity to big Capital Cities but a degree of isolation to this day, such as Canbelego and Graben Gullen. It is almost a delight, but I guess inconvenient, for those who live and work there to discover places off the beaten track within less than 100 kilometres of Sydney or Canberra. Go due west of Canberra via the aptly named Picadilly Circus gravel road junction, and you can vanish into the beauty of the relative wilderness of Brindabella, less than an hour from the nation's Parliament House and Cabinet Room.

To do all of this happily and safely it helps to have read the Tarboy columns and be reminded of the transport agonies and improvements of yesteryear. Blayney, on the western mainline and junction to Cowra and Young, deserves its special mention. It is in winter colder than cold for passengers changing trains. Mahoney, I am sure, would be delighted to know that today, against the trends and competitive odds, Blayney is enjoying a mini boom with rail freight operations (daily container freight trains directly connect Blayney to the Port of Botany) showing the right way forward for road/rail intermodal operations.

Even on the passenger front, in sharp contrast to the horse and buggy era, complete with the shortcuts and sidetracks explained so well by Mahoney, Blayney now has twice daily airconditioned XPT trains to Sydney and

Dubbo. Even the Indian Pacific passes through; it can be picked up nearby at Orange for journeys to Adelaide and Perth, and in 2004, with operational rail connections to Darwin. As we enjoy this progress, it greatly helps to keep a sense of balance and perspective by being reminded of so many colourful aspects of our heritage and especially our bush heritage.

We owe a debt of gratitude to James Arthur Mahoney, alias the "Tarboy" and his writings in the *Land* newspaper during the 1960s. These writings stand well with the work of many great columnists of yesteryear and today, such as Oxley in the *Land*, and also the *Bulletin* Balladeers like Henry Lawson, Breaker Morant and Banjo Paterson. Together, they help make our rich rural history come alive in a meaningful way for future Australians, wherever they live.

Tim Fischer
Joint author, *Outback Heroes*
December 2002

INTRODUCTION

"A Bushman Remembers" appeared regularly as a column in *The Land* newspaper from February 1961 until August 1962. It was written by James Arthur Mahoney, a distinguished agricultural journalist, under the pseudonym of Tarboy.

He wrote of a simpler time and of his memories of the old farming crafts, processes, habits and lifestyle. The Sydney Royal Easter Show was an annual pilgrimage for him. When some new piece of machinery came on the market, or he spotted, and inspected in great detail, the newest bit of equipment at the Show, he was likely to write about how the farm task for which it was designed was undertaken before mechanisation. Many of his descriptions of farm machinery, and how it was used, were about pieces of equipment that at the time he wrote stood neglected and rusting in corners of farms.

Work, and pride in doing it, is a consistent theme in the columns. So, too, is admiration for the old craftsmen. This is illustrated in a number of the pieces, especially those about the blacksmith's shop, coach painters, on tools, fencing and ploughing, and saddlers. And he wrote about bush life and entertainment.

The pieces are, of course, personal reminiscences about growing up and working on a small farm where everyone had allocated tasks. They describe the hard physical nature of farm work, the extraordinary multi-skilling that was necessary for the effective working of the property, and the simplicity of life in the bush.

Many of the columns mention horses and the work they did on the farm or in the wider community, providing power for pulling equipment or loads of farm produce, or for personal transport. He calls horses "grand creatures"—and one column praises the growth of Pony Clubs that would, in his view, help save the animals from "oblivion." These columns not only reflected the age in which he grew up, but also his abilities as a horseman. He was reputed to be an excellent horseman, and a breaker of some ability, although in one column he says this reputation was undeserved. At one stage his father raised trotters: ridden, not harnessed, and competed with them, often with Mahoney as jockey.

There is humour here and there, although it is not of the harsh, sometimes cynical, style of today. Like the topics, it is simple and some of it will resonate with those who have had similar bush experiences. Above all, the pieces are personal recollections of growing up and working in the bush, of the old ways of doing things, recollections re-told 40 years on as agriculture underwent one of its greatest periods of change, not the least of which was greater mechanisation.

The language is of a time before political correctness and gender-neutral writing. The columns were constructed in far more formal—indeed, sometimes quaint—newspaper style than would be the case today, Many words, for instance some for particular pieces of farm equipment, are not generally in modern use.

Mahoney was born at Canbelego, New South Wales, on 14 September 1906, the eldest of the 10 children of James Henry and Catherine Ethel (née Thoroughgood). He grew up on their mixed farm at Grabben Gullen near Goulburn, New South Wales. He had something of an education at Bannister Primary School, but had only one year of secondary education at Goulburn High School. He suffered scarlet fever and rheumatic fever as a teenager, the second damaging his heart.

His first experience in journalism was with the small news-sheet the *Corner's Voice*, published in Goulburn by Knowlman's department store. But he had been writing before that: in May 1923, aged 16, he was awarded a "Certificate of Merit" by John Fairfax and Sons Ltd for a short story with "a pleasingly original plot" which he had entered into *The Sydney Mail's* competition for children. In 1925, he joined the *Goulburn Evening Post* as a reporter, later becoming its Agricultural Editor. He stayed for five years until the Great Depression forced him to travel to Queensland looking for work. There he spent time on a dairy farm near Gympie, owned by his mother's brother, Arch Thoroughgood, and also continued to work as a journalist. When he returned from Queensland after the Depression, he married his fiancée, Ivy Scott, who lived next door to his parents who had moved to George Street, Goulburn. They had been engaged for the whole of his time away in Queensland.

He founded the Goulburn and District Tourist Bureau. In 1935 the Bureau, in association with the Goulburn City Council, published his book, *Goulburn, Queen City of the South*. At one stage he owned a small vegetable shop in Goulburn.

In 1934, he joined the reporting staff of *The Farmer & Settler*, becoming its second last Managing Editor in 1943, succeeding Gordon Bennett. He travelled extensively in New South Wales on assignment for *The Farmer & Settler*, was the paper's Dairy Editor and, before being appointed acting Editor in March 1943, was a sub-editor. By the time he took up the Editor's chair, the *Corner's Voice* had become a regular column, "The Corner", sponsored by Knowlmans in the *Goulburn Evening Post*. "The Corner" congratulated him on his appointment, recalling a story he had written for it about a magpie's nest that led the paper to urge him to a career in journalism. The magpie story appears in this volume. During World War II he worked as a casual sub-editor of *The Sunday Sun*, in Sydney.

He left *The Farmer & Settler* in 1953 after he bought *The Blue Mountains Courier*. Publishing this weekly fulfilled a desire to own a newspaper. In his early years, the *Courier* was printed at the Cumberland Press factory at Parramatta, but he later produced it on an old flatbed press installed in his backyard garage at Wentworth Falls. He moved his factory to premises in Froma Lane, Katoomba when he needed more room for job printing. Throughout the 1950s and early 1960s he was the Blue Mountains freelance correspondent for Sydney's *Daily Telegraph*. In 1959 he sold the *Courier* to Keith Newman, News Editor of Rupert Murdoch's then Sydney afternoon, *The Daily Mirror,* and took a position as a Sub-Editor on *The Land*. He travelled to *The Land's* office in Regent Street, Sydney, from his home in Wentworth Falls—a two-hours each way train trip. He worked on *The Land* until his death on 4 August 1962, just six weeks before his 56th birthday.

The rolled up copy of *The Land* and small bunch of wattle placed on his coffin by his brother Reg, symbolised his love of the bush and of his country and his 41 active years in newspapers, most of it in rural and agricultural journalism. He was an environmentalist, although he probably would not have recognised that description, and he supported conservation farming. One illustration of his concern for the environment was his agitation against the carving of faces of the explorers Blaxland, Wentworth and Lawson on a cliff face at Blackheath in the Blue Mountains. This was a one-man campaign that had him arguing against the proposal on an early ABC television current affairs program. He won the fight for, 40 years later, the carvings have still not been undertaken. Long before it became fashionable, he was interested in Australian history, especially of pioneering families, cities, towns and buildings. He was a long-time supporter of the Scout Movement in which he had been a District Commissioner. He was active in local affairs in the Blue Mountains, New South Wales; he was a member of Rotary and an Elder in the Presbyterian Church.

The last two columns written for the weekly editions of *The Land,* one published in the week he died, the other just after his funeral, have a touch of introspection, almost as though he knew he was dying and needed to reflect a little more on history and personal matters. He knew he was seriously ill but had told no-one. These columns appear as the last two pieces in the volume.

In collating the pieces, I have made some minor "sub-editorial" changes to the copy as it originally appeared. Some of the punctuation has been amended; some repetition has been edited out; sentences have been simplified; in some cases the newspaper style of publishing each sentence as a separate paragraph has been changed to combine linked sentences. Otherwise, the columns are as they were originally published, including the retention of the Imperial measurements of miles, yards and gallons and pints. The columns are not necessarily presented in chronological order; I have collated some in to groups of like subjects. For example, there is a series dealing with horses, another with timber cutting and a third with ploughing and harvesting. Similarly, there is a small series dealing with cooking, washing and tea making on the old farm, and another series about rabbits.

I had wanted to edit this volume before my mother died but never got around to starting it, perhaps because it had seemed like a simple task of locating, photocopying and re-keying the texts. No-one in the family had bothered to collect the columns as they appeared each week, and my mother would have appreciated a collated set even if it had not been printed and published. Now that it is done, it is just one record of the way things once were in rural Australia at a time when the nation relied almost exclusively on agriculture for its financial well being.

I hope that this collation of "A Bushman Remembers" helps Tarboy's grandchildren and great-grandchildren, who, of course, never knew him, and other young Australians, to understand their rich heritage and what farm life was like in the immediate post-Colonial period. Their generations are linked directly to his work through the cover design, which was produced in part by his great-granddaughter, Heidi Jagger.

I am indebted to *The Land* and its proprietor, Rural Press Pty Ltd, for granting me permission to produce this volume and for their kind permission to use the photographs from The Land Heritage Collection which decorate the chapters and front cover. The kindness of the General Manager Mike Harvey in allowing me to dip into the collection, in recommending CQU Press's Outback Books and in assisting me with access to hard copies of the newspaper, is greatly appreciated. So, too, is the help given to me to locate some of the columns by Ann Pender.

Sincere thanks to David Myers, publisher of CQU Press's Outback Books, for giving the volume a chance and for his many valuable and positive suggestions.

My wife, Janine, and daughters Emma and Holly and their husbands Mark and Alejandro, have been enormously supportive in my efforts to bring this volume together at a time of great change in our lives and I thank them for that.

James Scott Mahoney
Canberra, 2002

You name it, we've done it

My cousin Bill's voice could be heard all over the crowded bar as he regaled some of his heelers with highlights of our family history.

I was not in the party, but I could hear what was being said and I was surprised at Bill's knowledge of the family history, especially as he was regarded as the wild one, or no-hoper, of the modern vintage.

"It will be 175 years next year since the first member of our illustrious family came to Australia with Captain Arthur Phillip," he announced in a voice that made me flinch. "Don't ask me how he came—he could have been a convict, a soldier or a seaman who deserted, because at various times we have been all three.

"We have fought in all of England's wars, and sometimes on both sides. If you have ever read Kipling's poem, "The Irish Guards," you could well believe that it had been written about us.

"We have flogged and been flogged; we have been convicts, and we have been traps."

And he paused, thoughtfully, to down a schooner, as he recalled some of the more gloomy days of a fine old family.

"We have been sheep stealers and we have been squatters and sheep breeders," he declaimed. "We have duffed cattle, and we have bred cattle that have won prizes at Royal Shows and topped the export markets. But, and I'd have you mark this well, we have never had a horse thief among us. You've got to draw the line somewhere.

"We've always been in the advanced guard of those who've who pushed the frontiers further out and one of us actually got over the Blue Mountains before Blaxland, Wentworth and Lawson. Our man was in some sort of bother on the coast and reckoned that crossing the Mountains was his best chance of getting out of it. He seemed to do all right, because water, fish and other game were plentiful and it was not hard to get new clobber when other whites got across and it was not hard to mingle with them when they became more numerous. He got on well, too, because his local knowledge was valuable."

Bill borrowed a cigarette from one of his entranced listeners, who did not know whether he was telling the truth or a good story.

"We made fortunes on the goldfields and lost them later," he announced. "Some of us drove and guarded the gold coaches; other members of the family bailed 'em up and robbed 'em. As bushrangers we've robbed banks, and as police we've chased and captured bushrangers.

"We've built roads and driven coaches over them, we've piloted steamers down rivers and we've run camel teams in the Territory. We've laid railway tracks and driven trains and we've travelled in trains first-class as well as jumping the rattler.

"We've formed and directed companies and lived like millionaires and we've gone broke and taken to the track, but we've always been ready to come again.

"Our motto has always been that you might be down, but you are never out, and when you're down you've got a better chance in the bush than in the town, so the best thing to do is to get out and take a job—fencing, rabbiting, cutting wood, anything that will make more tucker. Even the young fellows in the family these days all get bush experience to help 'em later on if things get tough, which they are bound to do.

"We've shorn in the biggest sheds and we've been shearers' cooks, the best lurk in the sheds. We've raced horses to win and we've pulled them up. We've been bookmakers, registered and s.p. We've been publicans and we've been sinners. We've been parsons and as many of us have been criminals.

"In fact, you name it, we've been it.

"A couple of us are writers, journalists they call themselves but they are not up to the lurks of this family. With such a colourful history, they could make a fortune out of a book and a TV series, but they're too lazy or something to have a crack at it."

Wrong there, Bill, we've tried it, but without success.

The Land, 29 March 1962

We made a comeback with mead long ago

A news item from London says that mead, the truly traditional English drink, is making a comeback after five centuries of obscurity.

A former brewery chemist of Yorkshire, Douglas Morris, is sending bottles of mead, which is made from honey, all over Britain and abroad. Mead making, however, is only a sideline of his big bee-keeping business. He has 650 hives and made the first lot of mead "to see what it was like." When he found out how good it was, he began making mead for his friends and his sideline developed from that. He distils the mead from a secret recipe believed to have come from a monastery.

Our only venture into the production of mead was made with a recipe from an old cookery book that Mum owned.

We had a good supply of honey and Dad reckoned that it would be a good plan to convert it into mead, a drink he confessed that he had always wanted to try. He was a good hand at making hop beer, which could be given a terrific "kick" by the addition of various ingredients that were not supposed to be added for excise reasons, and he could produce a good wine from various fruits, such as elderberries, blackberries and even raisins. But mead was something entirely new and demanded a good deal of study of the recipe. This absorbed, it was more or less plain sailing.

I can't remember the details because I was young at the time, but everything seemed to go well. Whether or not the resulting product was anything like that produced by Morris of Yorkshire is another matter, but I remember that the corks in the bottles were tied down as a precaution because the mead was lively stuff when it was boiled.

Apart from water and honey, I can't remember the ingredients and, unfortunately, the recipe has been lost. Possibly, if we had been living in England, the recipe would have been handed down from father to son, with the necessary know-how. But Dad produced only the one brew, which he and his friends classed as a good drink, although a trifle heavy.

From my recollections, they merely toyed with the mead and did not quaff it down in large quantities as it was apparently handled in its hey-day in the old country. We youngsters were allowed a "taste," as with other hard liquor, and the memory is still good, even though it may not have been the traditional mead of old England.

Somehow or other a couple of bottles of that home-brewed mead was overlooked by the drinkers for more than a year. Somewhere in the corner the bottles had gathered dust and cobwebs until they were discovered by one of my younger brothers. He possibly had pleasant memories of his earlier taste of mead, so it did not take him long to extract the cork and put the bottle to his lips.

No one else knew what was going on, but there was some alarm when Mum found the young fellow asleep on the hearth and had difficulty in awakening him. He lurched to his feet without a word and staggered around the kitchen in what we thought was a faked condition of inebriation. But when we found the bottle half full of mead, we knew the condition was not faked.

I don't know what happened to the other bottle, but I have an idea it was treated with the respect it deserved.

Wonder if the Morris mead has a healthy kick?

The Land, 22 March 1962

That first ride in a motor car

Do you remember the first motor car you saw? I can remember mine—I had a bad toothache that morning, but the car made me forget all about it.

I was sitting on the woodheap at the back of the house when I heard the noise of a motor, drumming closer in the clear air of that frosty morning. Then the car came into sight—red in colour and with two men in the crew. They were well rugged up, as the car had no hood and was travelling at what was a fair speed for those days.

Our road was not in good condition and the car, more or less, jumped from one pothole to another. I have often thought that it would have been good to have known the make of the car, for it was well-built to have withstood the road on that journey. If they built cars of that make the same way today, they should be worth owning. By the time the car had passed our place, I had barely recovered from the wonder of it and realised later that I had seen little of the vehicle itself, despite the native curiosity of the small boy.

Some months later, however, we all had an opportunity of making a closer inspection of a motor car. It happened at the Sunday School picnic.

Some of the organisers had the bright idea of inviting a taxi-man from a nearby town to bring his car along for the afternoon. The idea was to charge sixpence a ride and share the take with the car's owner.

In those days, sixpence was a lot of money, but that car did not stop all the afternoon, except for an odd spell to enable the motor to cool down. It was then that we caught a glimpse of the motor, which seemed a tangle of wires and a complicated arrangement of tubes.

I've forgotten how many passengers made up a load, but memory, looking back over the years, indicates that it would be eight to 10 youngsters, a good cargo for the cars of the early days. However, packed in as we were, the thrill of a ride in a car was something that I, at least, have never forgotten—it was the highlight of my early life.

Those old veterans had much to endear them to us and the sleek models of today lack the same appeal.

Not so long ago I visited some friends I had not seen for years and when we were talking in one of the sheds I noticed one of the pioneer cars in a corner, under a load of old harness and other dunnage. It had been used as a fowl roost and had probably been the home of generations of mice, but the family who owned it would not part with it at any price.

There are probably dozens of old cars stowed away in odd corners on stations throughout the country, for many station owners were among the early motorists. One family kept every early car it owned and they are probably still to be found in the stalls of the old stables that were converted into garages. But those people are fond of the cars their grandparents drove, even though they may be neglected, and big offers from veteran car fans are spurned.

The Land, 13 July 1961

When snakes sent cold shivers down his back

Snakes always make a good subject for a yarn—and often for an argument.

Possibly because of my Irish ancestry, I do not like the slimy creatures at all and the sight of a snake will always send cold shivers down my back.

My first real snake scare came when I was about eight. It was early morning and I was on my usual job of running in the milkers. Against parental orders, I had gone out with bare feet, which helped to get the job over quickly as my speed increased without the impediment of footwear. On this particular morning I had the cows on the run—another parentally banned practice—and was making good time behind them, once I had them headed for home. My eyes were on the cows but that Irish intuition must have made me look down in mid-stride.

There, just where I was about to put my foot, was a plump looking black snake. Never an athlete, I must have made a new record leap for my age group to miss that Joe Blake, but miss him I did. Again came disobedience—I neglected to follow Dad's orders and kill every snake I saw and I was, for the rest of the day, shaken, as the saying goes.

Another morning, on my meanderings to school, I was testing out a neighbour's quinces for ripeness when a snake slithered away from my path. "Snake," I yelled, and the neighbour came running. His hatred of snakes was greater than his affection for his quinces and when I told him the snake had gone into a rabbit burrow, he got his double-barrelled shotgun and let both barrels go into the burrow. It was a careful man, and an even more careful boy, who dug that burrow out. There was not much of it, and even less of the snake, which had been shot to mincemeat.

People sometimes wonder why I don't take to freshwater swimming, but it all goes back to the day a cousin from Sydney was giving me swimming lessons in the dam at the bottom of the oat paddock. I was just mastering the dog paddle when I looked to the side. There, like Boy Charlton, a big snake (they're always big in retrospect) was making good time beside me. My cousin proved his skill as a lifesaver, but he could never get me to try my swimming ability in the dam again.

One of the funniest snake experiences in our family didn't happen to me, but to Dad. He'd been working hard in a dusty paddock on the morning of Christmas Eve and wanted a bath before taking the family to town for the afternoon. Our bathroom was a slab structure, tacked on to the side of the house, with plenty of ventilation between the slabs. There was no water laid on—you filled the bath with buckets. Dad was about to pour a bucket of water when a black snake (he was big, too) unwound himself from a spiral in the bottom of the bath and waved his head at Dad.

As usual, there was no good snake-killing stick around and I was too doubled up with mirth to find one quickly. "Joe Blake" left the bath and we nearly pulled the place down before we found him. It was a hasty search because no-one wanted to miss the Christmas Eve trip to town. When we poured a bucket of hot water down the hole we suspected the snake was occupying, he came out like a rocket—and he left his skin behind him.

The first column, *The Land,* 23 February, 1961

Blacksmith's shop was a home of skilled men

Y ou would travel a long way these days without seeing a blacksmith's shop, but even less than half a century ago there would be a smithy in every country settlement.

All blacksmiths did not work full time at their trade—some ran farms or shops, worked as carpenters, carriage builders and wheelwrights and more than a few were the local undertakers. But the coming of the motor car, the motor lorry and the farm tractor meant that the blacksmith's day was done. Fortunately for him, the blacksmith was a man with an open mind who accepted the changes as they came. Many a well-established motor garage of today had its origin in the local smithy—the blacksmith moved with the times and began servicing the early cars, in time putting in the petrol bowser

Being an adaptable and ingenious type, the blacksmith and his sons soon learned about the insides of the horseless carriages and became efficient, if not speedy, mechanics. They exchanged the leather aprons that were the regalia of their ancient craft for the overalls of the motor mechanic and the transition was barely apparent. Others retired or went into some other occupation, opening their almost neglected smithies when the need arose to provide service for someone in the district.

However, the passing of the blacksmith really began with the increasing sale of factory-made horseshoes, which farmers could shape and fit themselves with the use of a small portable forge, but that is another story.

We youngsters found those old blacksmiths' shops fascinating places and felt we were growing up when we were allowed to operate the lever of the bellows, supplying the air that turned the coals in the forge white-hot. When a big job was on, the blacksmith's offsider would pump that lever with vigour to get the right degree of heat in the forge and the smith himself would watch closely to make sure that it was right.

This was the application of science, but the blacksmith would have stared if you had called him a metallurgist. His knowledge had been acquired by long experience and was often developed from the advice and practice of his father and, possibly, from generations of craftsmen in the same family.

When the coals had attained the correct degree of heat, the smith would insert the metal to be worked, turning it and watching it closely as the heat changed its colour. It might be a horseshoe or a ploughshare, a drill or a pick being pointed, the tyre of a wagon being reduced in size, or any of the dozens of other jobs that came the way of the smith.

I was always impressed by the way in which the metal was dipped in a bucket of sand, as the heat intensified, to prevent it from burning. Carelessness in this regard could ruin an otherwise good job, with consequent waste of time and material.

When the heating process was completed, the smith, with his tongs, would place the piece of metal on the anvil, tapping it with a light hammer as, with alternate strikes, his offsider, or striker, would yield a heavier hammer. The "tap, tap, tap" of the two hammers provided a form of music that has been woven into great works by master composers. As the job took shape, the use of the hammers became more delicate until the completed state was reached. Then came the tempering process—the metal was heated to a certain shade of colour and then dipped into a bucket of water to get the correct temper. With ploughshares and such tools as drills and picks, the points would be the last section to be immersed in the water, the reason for this being that the hardness of the temper ran out at the point.

Some blacksmiths were true artists and their work in wrought iron would fetch high prices in these days when the thoughts of professional decorators are running in those directions.

Although it was inevitable, the passing of the blacksmith is regretted by many. There is not so much romance in running your car into a garage for service as when you rode the saddlehorse to the smithy and drove or led the farm team there to have them all shod, sometimes taking the whole day about it.

The Land, 25 May, 1961

On and off the handle

Handles are always a cause of trouble about most places.

You have the axe handle on which it is difficult to keep a head no matter what care you give to the wedging, or the handle that breaks at a most inopportune time. Sometimes it is difficult to get a satisfactory handle for the claw hammer, but, on the other hand, hammer handles will last for years although the tool is used no differently. Pick and mattock handles can be a constant source of trouble, especially if they have been made from timber that has a tendency to cast off minute splinters—and all suppliers are not at all careful about that kind of thing.

Talking of pick handles reminds me of a Civil Defence Organisation meeting I attended not so long ago. The talk was of possible panic among refugees and one of the leaders said the best way of quelling such outbreaks was with a supply of good pick handles, wielded by experts. We could see ourselves being issued with pick handles, but the non-existence of any funds for our CDO may mean that we have to quell panic in some other way.

Pick handles are better than baseball bats for bringing people to their senses, as I saw during a near-riot between Australian and American troops in a northern town during the last war. It was not a national fight—there were Yanks and Diggers on each side, and when the Military Police of both nations arrived, they went into action as a combined force. The American MPs swung their baseball bats and the Aussies their pick handles without any bias of race, colour or creed, and the little upset was soon over. An inspection of some of the casualties revealed that the pick handles had done by far the most effective job.

There are, of course, other uses for pick handles.

An inferior or faulty handle can mean a lot of trouble, especially if there is a big area of burrs to cut and the weather is such as to encourage the seeds to ripen early. You need a smooth, strong hoe handle, with a certain amount of "whip" to do a good job, and if you are cutting burrs from horseback, you will bore a hole in the end of the handle so a thong can be attached. This means that the hoe can't get away from you just as you keep control of a polo stick with the use of a similar thong around the wrist.

My experience was that short shovel handles always gave better service than the long type, for some reason. Possibly, the long handle had to stand a greater degree of strain, especially when you spent a lot of time leaning on it.

Handles of most tools generally stay on all right in damp weather, but with the arrival of summer they have a habit of flying off at the most inopportune times.

The Land, 1 February 1962

Flying off the handle can have its dangers

Tools have a habit of flying off their handles during the hot weather of summer, sometimes with disconcerting results to the user.

It is no fun to be wielding an axe or a sledgehammer and for the tool to become divorced from its handle, but that has been known to happen. One immediate effect is for the user to continue the blow and, without the weight of the tool, he is thrown off balance, sometimes with dire results. I've known people to suffer serious injury from such incidents through toppling over the log they have been cutting, or falling into the rocks where the sledge hammer has been used to power a drill.

Looseness of tools on the handle is dangerous. It can be countered to some extent by leaving them in a tin of water overnight, when the handles will swell into the heads. But care has to be exercised during the day to ensure that they are not left lying about in the sun. At any rate, the soaking treatment does not seem to last the whole of a farm working day, which was much more than eight hours in my young days.

Some people care for their tools in a devoted way, and they treat the wooden handles in the same way that a Test cricketer looks after his bat. The handles are oiled, kept free of dust and the tools all have their own special places in a tool shed or workshop.

We never did things that way—you'd find our tools anywhere at any time and finding them meant that the start of a job was delayed. Perhaps that was the reason we were always in trouble with the handles.

My experience has been that the greatest trouble with handles comes from the homemade variety. Some people seem to have the knack of being able to turn out a satisfactory "do it yourself" article, but they are rare. Generally, the homemade variety is roughly shaped, ill fitting and, because of unsuitable material, apt to splinter and break.

I have, however, seen axe handles made by old bushmen that have lasted for years, but they are really a painstakingly efficient job. Great care was taken in the selection of the timber, which was lovingly shaped, smoothed and fitted to make a handle as good as any that could be bought.

Against that, you would have the rough hammer or tomahawk handle, hewn from a piece of wattle or blue gum, with no attention paid to the way in which the grain went. Such handles did not last long. If the tools were used much, the owner would, in desperation, do what he should have done in the first place—buy a new handle.

The Land, 8 February 1962

Splitting posts was job that made call on precision

When I was a lad, a maul and wedges could be counted as almost standard equipment on every farm.

There was so much fencing and building to do, and so much timber available, that the maul and wedges could have been put almost into constant operation. And if you did not own the tools yourself, you could usually borrow them from a neighbour.

We'd fell a tree with axe and crosscut saw, then saw it into appropriate lengths for fence posts, which we would proceed to split.

If the grain was straight and clean, splitting was not a difficult job, but if it happened to be curly or woolly there would be trouble for the splitter. It was the first split that counted most—if it was straight and divided the log in half, you were well on the way and it would not take long to split the halves again, and then again until the desired size of the post was reached. But if any difficulties were encountered, it would be necessary to change tactics and to improvise.

Sometimes more wedges would be needed, and the way this was overcome was to cut a piece of green timber to the required shape. After the split had started, this piece of timber was driven in to keep the cleavage open and was generally followed by another piece at the other end. Then the steel wedges could be used to keep the division increasing, more or less playing a holding game with the wooden aids.

Splitting timber was not always a safe pursuit as wedges were likely to be squeezed out of their position in the sappy wood and to be flung into the air as if they had been released from a catapult. To be in the way of a flying wedge was not one's idea of a pleasant interlude.

Post-splitting took a degree of skill, but the really fussy job was splitting droppers. A poor splitter would have droppers varying in width and depth, heavy and ungainly, the type that did not make a good fence, either from the angles of appearance or durability. A good splitter, however, could produce droppers almost as fine as sawn palings, light, and yet strong enough to do their job, even in size and, verily, a joy to behold. He would not rush into his job, however, but would have his wedges overhauled and sharpened by the blacksmith, and he'd make sure that the maul was in good condition. Generally, he would make great use of the smaller wedges and if he was a professional, he'd probably have a separate set especially for droppers. Of course, he'd follow the same procedure when splitting palings, the only variation being in size.

Some of the best splitters were old-timers who had gained their skill splitting shingles. I have never seen a splitter at work on shingles, but it must have been a craftsman-like job.

Terrific tales used to be told about accidents that befell splitters on their own in the bush.

One of the most gruesome was about the character who had opened a good division and had decided to turn his log half over. Using both hands, he gave it a push, only to have the wedges fly out and the closing gap imprison his hands. That was a really helpless position, especially as his billy-can and food bag were just beyond his reach. As the story goes, he perished there and his skeleton was found many years later by a new settler cleaning up the bushland.

Sometimes the tale was varied and the unfortunate victim became a bushranger, held with his hands in the cleft until the police arrived. But this never impressed me. In fact, I did not believe it because a bushranger would have far more sense than to get into such a predicament by dropping his gun to put his hands in a place where there would be some difficulty in getting them out if anything went wrong. I may have credited the bushrangers with more sense than they really possessed but I still subscribe to the same view.

The Land, 21 September 1961

Post and rail fences were well-built jobs

Do you remember the old post and rail fences that used to mark out boundaries?

They were popular before wire netting came into favour, when they served all necessary uses, especially in keeping stock where they should be kept. But their popularity declined when the rabbit increased because two, three or four rails offered no barrier to the furry pest. Some landholders draped wire netting over the rails in an effort to cope with the rabbits, but this never proved really satisfactory and all of those fine old fences eventually became firewood.

They were well built from good material and a lot of work went into them.

The posts were solid and were morticed to take the rails. Posts were sized to avoid any great discrepancy when they were erected, as one of the objectives was to preserve a certain amount of uniformity. Rails were split in the same way as slabs and were adzed to bring them to something about the same size and to remove any splinters. Ends were shaped to ensure snug fits into the mortices of the posts and some of the more careful owners, to avoid damage by white ants, would have the ends of the rails dipped in hot tar before they were fitted.

Another step taken by the fussy types was to bolt or nail the rails into the posts to prevent any movement, but this was done when the sap had dried out and the fence had "settled down." It was not a general practice, however, and most of the fences would rattle when stock scratched themselves against the fence. Some of the more prosperous owners would paint the post and rail fences in front of their homesteads and along the drives or entrance lanes, but this was regarded by most as being unnecessary "show" and was done only by those who were sure of their position in the community. Another practice was to string a couple of strands of wire along the fence between the rails. This added strength to the structure.

One of the most popular uses of the fences, especially near towns and other settlements, was for advertising. Various businesses and services would paint their messages on the rails to attract the attention of the passing traffic and sometimes the messages would have a bit of humour. Mostly, however, they confined themselves to the bald announcement that Joe Blow was a good blacksmith and that Tom Smith had a grocery store in Main Street. A revolutionary type would perhaps offer such advice as "Chew Chew's Bread on Tuesday" or "Visit Paddy Murphy's Pub for the Best of Cheer" when everyone knew that the main feature of Murphy's bar was its arguments and fights.

I have often thought that those old post and rail fences would have been a godsend to the modern advertising man who would probably have been paying the owners big money for the "space" they could make available. It is significant that the departure of the fences brought the big and ugly hoardings to much of the countryside as advertising entered a new phase. Apart from the advertising, which was only incidental, some of our gracious living disappeared with the old post and rail fences.

The Land, 2 November 1961

Gates present some problems

Farm and station gates have always been an interesting subject, especially to those who have been forced to open possibly hundreds of them on a day-long trip.

Despite the growing popularity of ramps or grids, which avoid any opening at all, there are still many gates on country properties and the variety is as extensive as ever it was.

Possibly the first type of "gate" used in the early days was the sliprail. This could involve anything from one to half a dozen rails, of differing lengths and weights, and a person did not need to be in a hurry when he came to a set of slip rails. It took some time to take the rails down and just as long to replace them.

Some of the early settlers, however, believed in solidity and they constructed impressive gates of hardwood, heavy contraptions that led to all sorts of trouble unless they were well swung on massive and deeply set posts.

Others went in for gates of somewhat similar pattern, but constructed of saplings and therefore lighter and easier to open and shut. These, however, had their disadvantages, particularly because of their lightness. They soon went out of shape if scraped by a wheel hub or if some of the wilder horses and cattle endeavoured, without success, to jump them.

Another thing to watch was to make sure that the catches, or latches, of the gates were correctly positioned when closed. Otherwise, the gate would swing open, stock would get out of the paddock and the swinging gate, bumping against the post and the fence was soon in need of repair. Unfortunately, time could not be spared for much repair work and the gate continued to deteriorate until it became a hard to handle mass of poles and wire that was discarded, long after, when it became unmanageable.

Much better were the gates constructed of 3 x 1 or 4 x 1 hardwood, bolted together. If the nuts were kept tight and the balance of the swing retained, those gates were easier to open and shut and lasted for many years—if no one ran into them or scraped them with wheels.

The nightmare of all was the "German gate," constructed of wire, including the barbed type, strung to a round dropper each end and, generally, with another dropper in the centre. I've been tangled up in "German gates," which then resembled barbed wire entanglements, and I've wasted large amounts of time trying to close the gate and pull it taut with the dropper on the free end. It was fatal to endeavour to open one of these gates from horseback, as the wire would scare the horse and its dancing would make the task so difficult that you might even be pulled from the saddle into the gate.

When new, the "German gates" were bad enough but after they had been in use for a time, and had had "repairs" made to them, they were definitely the last word as time wasters and arousers of tempers. There are still "German gates" on some properties, probably to scare off unwelcome visitors.

The first break-through to really efficient and long lasting gates were the Cyclone type made from tube steel, well braced and carrying a netting barrier against the rabbits. Farmers and graziers who took pride in their places invested in these gates, which improved the efficiency of operations, had a beneficial effect on tempers and reduced the time spent on maintenance.

Today, the efficient landholder builds ramps, which enable vehicles to pass over but keep stock in their correct paddocks. However, the whole story of the Australian gate can still be seen in many parts of the country—from the sliprails forward.

The Land, 8 June 1961

Expert axeman always keeps blade in order

One of the most essential pieces of equipment on any property is a good axe.

An axe has many uses, from chopping wood for the home to felling and shaping saplings for buildings, as well as assisting in the cutting of any millable timber.

Where axes are regarded highly, you'll always find a grindstone and an oilstone. An owner who appreciates the value of his axe will never touch it with a file—when sharpening it, he will grind it carefully and then add the finishing touches with the oilstone. He'll take care how he uses the axe and the type of timber he sinks it into. If the timber is rough, too knotty or liable to have nails or spikes in it, the good axe is never used, but an older, more knockabout type, gets the job.

It has been said that some enthusiasts lavish more care on their axes than they do on their wives and families. An axeman who is a real fanatic will not let his wife and family touch the "good" axe, which has its own special place in the tool shed. However, the second favourite axe does not get nearly so much care. It is kept sharp enough to cut a toe off if used carelessly and is adequate for most chopping jobs about the place. It is the axe that is used for killing the Christmas fowl or turkey and for drastic pruning of the fruit trees that have been allowed to go wild. Anyone can use the utility axe, which means that it uses up quite a lot of handles, especially where young folk are concerned.

Replacing a handle is not always an easy job. Sometimes the old butt comes out of the head as easily as anything but other times it is reluctant and that means a lot of chiselling and cutting, which is not facilitated if an iron or steel wedge has been used. However, patience does the trick, but if you want to be really nasty to an axe lover, burn the butt out by placing the head in a fire. This, you'll be told, destroys the temper of the axe steel and really means the end of the axe.

Everything, of course, depends on the temper of the axe steel, which enables it to be brought to a fine cutting edge and to keep that edge until it has been sharpened away.

Tomahawks come into a separate category, but when I was a lad half-axes were popular. These were smaller, lighter than the full-size axe and were apparently made specially for growing lads who found the big axe too tiresome to swing for any length of time. It has been said that the half axe was made originally for the women of the family who could keep the woodbox filled with a lighter implement.

The Land, 28 December 1961

Woodheap duty was a regular daily job

Firewood is becoming a problem in some of the older settled districts of Australia as the trees have been destroyed and not replanted. This means that wood has to be carted long distances, or some change made in methods of heating and cooking.

Unless a wood lot is developed on every holding, the use of wood for fires on farms and stations will become a thing of the past—as in the towns and cities. But there was no shortage when I was a boy—in fact, the general idea was to get rid of the timber to make way for crops and herds.

There was plenty of wood within a few yards of the house at first, but that was eliminated through Mum's desire to have the place cleaned up so she could get a garden going.

Gradually the wooded area was pushed outwards, except for the shelter belt of green timber. In time, it became necessary to cart wood to the house. The distance was not great, but it was easier to get a load or two in the dray rather than carry it all.

Our woodheap loomed large in my mind because it represented one of my daily jobs. I had to cut and split the wood and fill the woodbox to ensure a constant supply for the stove. Winter brought the heaviest work, as there had to be enough wood for the open fire as well and even the use of the axe in the brisk and frosty air was not sufficient recompense for the work. There were times, I fear, when I slummed the job and someone else had to make up for my idleness through the day.

Younger members of the family had the job of maintaining the supply of kindling—light sticks and "candle bark" gathered from the shelter belt. That material was always falling and it was merely a matter of picking up an armful and carrying it to the side of the woodbox, where it remained in readiness for next day's fire lighting. However, the kindling crew frequently forgot its obligations and it was nothing to see a youngster make a dash for the shelter belt some cold morning when Dad got up to light the fire. His idea was that the neglectful one would remember his obligations if he was dragged from the warm comfort of the blankets and made to collect the kindling. Of course, I suffered the same treatment, but as I grew older I realised there was no future in it, so made sure that there was enough wood in to at least get the fires going in the morning.

This wood getting, fortunately, was not a constant job. Dad, who liked to wield the axe, would hop in and cut a supply at times and we sometimes had visitors who'd put in a bit of time on the woodheap.

Our woodheap became a centre of social activity, too. There would always be a couple of blocks too solid to split; they made useful seats where upwards of half a dozen people would sit and yarn, especially on a sunny Sunday. As they yarned, some would whittle sticks, others would poke the toes of their boots into the chips and down to the rich compost-like soil that had developed from the rotting chips.

As we grew older and began to go farther afield, the woodheap gained some importance in courtship. It was far enough from the house to give privacy, yet close enough to allow the young lady's family to keep an eye on us: courtship in those days was not an unchaperoned extravaganza.

The Land, 22 June 1961

Mum will need a lighter saw

A new concept in getting the family wood supply will have to be evolved with the rapid growth in popularity of the chain saw.

When you see axemen of world renown using and advocating chain saws, you realise that the day of the good old axe is nearing its end, except for certain light work about the place. And no longer does the circular saw have its place around the homestead—everything it did the chain saw can do, and generally do better.

I have seen men fell big trees and cut them up where they stood in a fraction of the time they would occupy if they used axes and crosscut saws, with the accompanying mauls, wedges and other gear that is constantly getting lost. On these jobs, the only use given the axe is to remove the lighter branches from the main trunk, all heavier stuff being attacked by the chain saw.

On a recent vacation, I was fascinated by the speed with which three men cleared a block of land on which the builders were already starting work. Several big trees had to come down, and they were felled in the right direction to prevent any damage to anything on the block. Then the men hopped into them with their saws, trimmed the branches from them and cut the big stuff into lengths that could be handled easily—and they did not appear to be working hard.

They cleared that block in less than a day, working to the music of the chain saws ripping through the timber and the occasional thud of an axe as the smaller branches were removed.

This job was of particular interest because the men had cleared three or four trees from around a house I had bought in a Sydney suburb. They were at it two days, which included cutting up the wood with axes, although they were no mugs, having chopped in top events at Royal Shows. At times they used crosscut saws to cut the trunks into smaller lengths and although it provides some of the poetry of motion, this sawing can be a laborious business, even for a well-matched pair of sawyers.

Possibly few other jobs bring out the perspiration as operating the old crosscut does, and, unless you are perfectly fit, you are always ready to crawl into bed after a day on one end of the saw.

But now the chain saw has ended all that. It has brought a new conception of his work to the timbergetter and to those who have to supply wood for the homestead, whether for the cooking stoves or for winter heat. Logs can be cut into billets of any size with the chain saw, apparently with ease. Now, I suppose, there will be a call for different sizes of saws.

Axes used to come in three sizes—the big one for Dad, the little one for the sub-teenager lad, and, of course, the light one for Mum, who needed something a bit better than what could be classed as a toy, but not as heavy as Dad's axe.

Will we get three sizes of chain saws, as all the models I have seen appear to be a trifle heavy for a woman to handle on the woodheap?

The Land, 3 May 1962

Where willows weep

There is nothing more pleasant on a hot summer day than to see and hear the water of an inland river gurgling and trickling over the clean sand and gravel. If that section of the stream is shaded by willows, it is even more delightful and it becomes a temptation to forget what one is doing and sit in the shade, possibly dangling bare feet in the water.

Maybe you are on the edge of a pool deep enough for swimming and there is nothing better than a dip in the crystal clear, fresh water streams of the inland. They make even the most famous surf beaches second rate. There is no sticky feeling from the salt, no danger from sharks, jostling surfers and the assortment of balls, rubber surf rafts and other hazards the beaches present.

Any river has its attractions, except when it is swollen by flood, but those along which willows grow are by far the most attractive.

In the district where I spent my school days, two types of willows grew along the streams, the weeping willow and the basket willow. Basket willows produced boughs that were straight and we were told that they were used for making baskets.

An intriguing story about the weeping willows was that the first slip had been brought from St. Helena, where Napoleon spent his exile, and they were known as "Napoleon willows." The story was that an early settler was a Frenchman who had been an officer in Napoleon's *Grande Armée* and that he had collected the willow slip when his ship called at St. Helena on the way to Australia. Settling on the bank of a river, his first action was to plant that slip, which thrived. Floods and stock broke branches from the original tree and they floated downstream until they lodged against the bank where they took root. It was not many years before willows were growing along the whole stretch of the stream.

Droughts were not common in our district, but when they did come the willows proved a godsend to many as emergency fodder for their stock. They lopped the branches, dragged them away from the banks and the stock went to work on them, consuming leaves and the thinner branches. There could have been little nutriment in that fodder, but it provided bulk at least and played no small part in keeping large numbers of stock alive.

Another attraction of the willows was the use of the dry wood to get a camp fire going. It burned rapidly, but it gave off a pleasant aroma and added its own distinctive flavour to the tea.

Even the hot work of droving a mob of stubborn sheep could be forgotten in the pleasure of a meal taken under such circumstances—the gurgling stream, the cool shade of the willows, the distinctive odour of the smoke, the tea with the willow wood flavour.

The Land, 7 December 1961

Wattle trees have had many uses

Wattle trees were once a valuable "crop" on many small holdings and they reached productive stage without much human aid. In fact, that was one of their advantages—other jobs could be done as the wattles were growing.

Their greatest use was the production of bark for tanning. The bark was stripped from the trunks, allowed to dry, and it was then bundled up and carted to the tannery or to the local dealer, who bought anything he could re-sell at a profit. Another use for the wattle was as wood for bakers' ovens. It provided the greatest heat available for specialised baking in the days when ovens were wood fired and it would always bring a good price. For wood, however, the wattle had to grow to a greater size than when it was stripped for tan bark.

Trees were stripped for their bark when they were mere saplings, apparently because the strippers happened to be in need of money. After a clump of trees had been stripped and the wood dried out, the knowledgeable owner would run a fire through the remaining trash. This speeded up the germination of seeds and the clump would soon regenerate, possibly spreading a little around the edges to provide more saplings for later stripping.

More than one farmer fed himself and his family between cheques for normal crops and wool by capitalising on the wattles that grew on his property. It was common to pass dray-loads of wattle bark and wattle wood on the way to town as advantage was taken of the natural harvest.

Stripping wattle bark was a job that demanded a sharp axe and a certain amount of skill. The most unpleasant part of it was the way in which the sap stained the hands and the clothes of the worker.

An excellent tanning solution could be prepared from wattle bark, even for "do-it-yourself" jobs with odd hides on the home farm. Visits to the local tannery revealed sheds full of the bark and shredded bark going into the evil smelling tan pits. That particular tannery, a section of a boot factory that is among the oldest continuing country industries, made its reputation more than a century ago. Wattle bark, I understand, is still used for tanning the leather of top grade footwear that is in wide demand, although chemicals are used to tan the cheaper grades.

When I was a boy, some consternation was caused among the wattle bark producers when they learned that South Africans had acquired the seeds of Australian wattles, cultivated them and produced so much wattle bark that they could export it to Australian tanneries. Even when this threat became a reality, none of the bark producers bothered to try similar methods—they were still prepared to let nature do the work.

Another use for wattles was outlined to me by an old-timer who was interested in tree-planting. He claimed that the wattle, being a legume, would be invaluable as a pioneer plant that would supply nitrogen to the soil for later plantings. He suggested that wattles should be grown on sites for future shelter belts and, when they had matured sufficiently, they should be stripped for their bark. Then the permanent species should be planted in the shelter belt. My friend claimed the trees would thrive because of the excellent conditioning given the soil by the leguminous wattles. Sounds a good idea, but I haven't heard of anyone putting it into practice.

The Land, 18 January 1962

Tackle the job properly and you can achieve much

Some of the old-time Merino flocks in a district of small farms I remember would hardly pass muster as good woolcutters today. Few people bothered to weigh the individual fleeces, but they could hardly have exceeded 2 or 3lb at the most. Those sheep were small in frame, wrinkled and possessed most of the type faults that the experts are dead against today.

If a "gun" shearer got on to any of them, which was infrequent, he would ask one of the shed hands to grab the other end so they could be pulled out like a concertina. This facetiousness was, of course, a typical comment on the wrinkles, which the owners put down to the "Vermont influence." Apparently, some studmasters had introduced sheep from Vermont (USA) into their flocks, the main feature of the introductions being their intensively wrinkled state.

It was, however, drawing the long bow to blame the "Vermont influence" for the wrinkled sheep I remember as a boy. It was possibly due to the mixture of types and strains from which the flocks were evolved, without any culling or classing at all.

Those sheep might have cut small fleeces, but the wool was of fine quality and always sold well because of that, although the clips lacked any sort of preparation. The only treatment the wool received was the removal of the dags prior to the remainder being bundled up and thrust into bags or butts. Those woolmen would have been astonished if anyone had introduced them to a standard woolpack.

Shearing was done by the owner with the blades, but if he could not shear, the wool was taken off by a neighbour who, at some time or other, had worked in a shearing shed or who had inherited the craft from his father. Shearing was done in the hayshed or under a tree and the most important equipment, next to the shears, was a good supply of Stockholm tar. This was necessary because the wrinkles meant that cuts were many and some of the sheep seemed to have been showered with tar when the shearer let them go.

I have seen the descendants of John Macarthur's flock at Camden Park and, although they are a long way from our big cutting Merinos of today, they are of even better type than some of the farm flocks I remember, especially where size is concerned. However, one cannot be too hard on those old flockmasters, as their sheep represented only a sideline and any returns from them were straight-out profits.

In time, constant propaganda and the work of neighbouring owners brought improvement, slow though it might have been. More attention was paid to plain-bodied sheep and by the expenditure of a few pounds on a better type of ram, size was introduced into the flock. Then came some attempt at culling and the flock gradually improved, especially when the son of the owner became interested in the sheep.

Some of the farmers changed over to crossbreds for fat lambs, but those who stuck to Merinos and worked to improve them, eventually reaped their reward.

Nowadays the descendants of those early Merinos are producing wool that sells at a high price, and other breeders seek the rams. It just goes to show that you can achieve a lot from unpromising material if you tackle the job seriously.

The Land, 11 January 1962

Ploughing a straight furrow is an art

There should be little difficulty in ploughing a straight furrow in these days of mechanised farming. Seems to be a case of setting off on the tractor and more or less locking the wheel in position so it will steer a straight course. And farmers still like to have their furrows straight, even in this jet age, judging by what one sees in paddocks throughout the country.

It was not always easy to plough a straight furrow in the days when the horse was king. Every horse, whether worked singly or in a team, had its own idea of where it wanted to go and that idea did not always coincide with the wishes of the driver. Hence, ploughing a straight furrow was an art.

Everything depended on the first furrow ploughed in each "land" or section of the paddock. If that furrow was straight, the rest of the "land"—and usually all the paddock—would have a pattern of straight furrows.

With the careful ploughman, there would be the stepping out at right angles to the boundary fence. At the far end of the paddock, a stake with a white marker would be pushed into the ground to indicate the correct distance. At the starting end, a mark would be made on the ground and the ploughshare would sink in that exact spot. Then came the real test of the ploughman's eye and skill. The team would be started and driven at a steady pace towards the marker, usually with the driver sighting along the neck and head of the most conveniently placed horse. With a good steady team, soil in the best condition and a ploughman with a true eye, that first furrow would be straight—not perfectly so, but with curves that could rectified as the ploughing proceeded.

Such methods, however, were not for the true artist. He would have a light, single furrow mouldboard plough, to which was harnessed the quietest and most reliable horse on the farm. Using the marker system, he would set off, going as straight as a craft with an accomplished navigator. That was "striking out." Once the first furrow was ploughed, the team with the two or three furrow plough would go into operation, throwing the next sod into the first furrow. All was not plain sailing afterwards, however. The team had to be kept in line to ensure that the succeeding furrows were straight, as a step off course or a moment's carelessness on the part of the ploughman could mean crooked furrows. With one "land" finished, the procedure was repeated until the job was completed, with headlands on the four sides of the paddock exactly the same distance from the fences. It was a work of art and the man who had row upon row of dead straight furrows was a man to admire.

When we became lazier and used ploughs which had seats so we would not have to trudge behind, some time elapsed before driver and team worked out a system that ensured straight furrows.

Most of the ploughmen in our little district were straight furrow men, but one or two could see no reason why a farmer should be so fussy. But the "straight" men regarded such heresy as an indication that those who felt that way had no pride in their work.

Too, there was the sad case of Old Paddy, a man beyond redemption. Paddy was a reasonably good ploughman—after he had signed the pledge. He did that for regular periods of 12 months every year, and after each pledged period ran out, Paddy would indulge in a mighty spree that would last for weeks and until the family and the priest got him to sign the pledge again. Usually at some time during the spree period, Paddy would decide to do some ploughing. Under normal conditions his horses were quiet and easily handled, but when Paddy "had drink taken", as his brother put it, the team's temperament changed. With horses a bit flighty it is not easy to plough a straight furrow, even when you are cold sober. But in Paddy's condition, with nervous horses, his furrows were as crooked as a dog's hind leg. Even Pat himself was ashamed of them after he had sobered up and he'd hasten to plough those particular few "lands" again.

I often wonder how Paddy would have piloted a tractor after he had partaken of the "craythur."

The Land, 16 March 1961

Speed the plough was Dad's motto—and he got results

Automation had an early introduction to our farm, as it arrived in the horse era, preceding the tractor by some years.

Dad was always proud of the speed of his work teams, which enabled him to get his jobs done in less time than it took his neighbours with their heavier steeds. He accomplished this initially by concentrating on lighter horses, whose main virtue was speed, and the farm plant, one could say, was built around them. "Light draughts" was the term applied to our horses, whose walk approached the speed of a trot. It took some time to get together a team of uniform speed and somewhat alike in size, because uniformity was necessary to avoid discrepancies in harness, and so on.

In those days few ploughs had seats. Mostly the driver, the harrows, scufflers, cultivators and so on, followed them. Seed drills had a platform at the rear on which the operator stood; the reaper and binder was equipped with a seat. Our faster horses, however, meant that some adjustments had to be made to these machines to ensure their efficient operation because in those days they were built for horses of moderate speed. Anyone who has had experience with precision machinery will know that speeding up the operation demands some adjustment, but once it is made the job is far more efficient than at the slower rate, all of which favoured our faster team.

No one liked harrowing because it meant walking over the ploughed land behind the horses and it was impossible to escape some dust, no matter on which side of the harrows one walked. Dad made experiments with small carts towed by the harrows to provide a standing place or a seat for the driver, but when he was almost upended into a set of diamond harrows, he decided that some other method would have to be adopted. What was more natural, then, than to drive the harrow team from a saddle pony? And so we were trained to ride the harrows, controlling a team of four draughts with a set of reins and guiding the pony with our knees, because we had little chance of using the bridle effectively when direction had to change.

After some initial difficulties, all went well with the new method of harrowing. Of course, we collected a fair issue of dust and had to counter the difficulty of the team having an idea that it was engaged in a race, which meant that the harrows jerked up, and sometimes over, and a great deal of the paddock was missed. One of the worst problems for a time was the difficulty of getting the pony and the team to work in unison and some of us were pulled off and into the dirt on occasions. But as the method was developed and brought near perfection by use, even the girls wanted to have a crack at it. Being good horsemen, they soon caught on and, truth to tell, showed the rest of us up because of their superior horse control. They stuck at the job, too, and their eagerness for the new method of harrowing released the man and boy power for other jobs, all of which could be put down to the system of automation.

Came the time when a new plough had to be bought and everyone pressed for a unit complete with seat. Not that Dad needed any pressing—it had been his intention all along to get a plough with a seat that he reckoned would cut a lot of time off the job. His preference was for disc ploughs and it was a proud day when he brought the new three-furrow job home, all bright with its red and green paint. There were no problems at all with this plough in the older paddocks, which had been well cleared of under-surface hazards. However, it fell to my lot to strike the first snag—literally.

One afternoon, I was ploughing a paddock not far from the house and the team was moving along with its usual accustomed briskness, keeping the correct line, turning the right way and actually doing the job. Halfway along the furrow on one trip there was a loud bang as the rear disc struck a lump of quartz, unobservable from the surface. The impact snapped the springy steel of the seat standard like a carrot. By that time, the plough had passed over the quartz, but I hit the lump full force with the back of my head and was almost knocked out. It was a groggy young ploughman who eventually stopped the team and went for assistance. After some acrimonious discussion, we rigged up the seat with U-bolts and got our ploughing on the way again.

Our fast horses branded us among the neighbours as a family afflicted with some degree of madness, and accidents like the plough seat episode, when they heard about it, provided confirmatory evidence. But over the years, one noticed, there was a trend to get away from the slower type of heavy horses on all neighbouring farms and to go in for the lighter, more speedy type, so the madness must have been infectious.

The last column written. *Land Annual,* 17 October 1962

Farm dogs were a motley crew

Farm dogs come in all shapes, sizes, colours, breeds and blends. They range from the savage hound that you have to keep chained up, except when he is working, to the fat and indolent mongrel you can't kick out of your way.

In my young days every farm had at least one dog, but few of them were any good for anything, except getting in the way and eating eggs.

Some people, of course, had well-bred Kelpies and cattle dogs, but there was insufficient work for either on the average small farm and the poor dogs would almost die of boredom. The Kelpies would "work" the fowls until they had them thin and despondent and the cattle dogs, cheated of the job for which they were born, would keep in touch by "heeling" any children they could find. Some would be so eager for work that they would start heeling the horses, frequently a fatal pursuit for a dog that had not been trained to do it. Horses have a different method of kicking and, being unaware of this, the brains of many an otherwise good cattle dog have been scattered about the yard because he did not drop to the ground quickly enough.

Not having enough work, the sheep and cattle dogs would be out of form when it was necessary for them to revert to the occupations for which their breeds were evolved and they would often be more nuisance than help. Generally, however, the farm dog had little to link him with any known breed, as he was the product of numerous crosses, sometimes carrying the strain of a wide ranging paternal ancestor possessed of great fighting ability when it was needed.

Dogs would spend a lot of their time in the shade—under a tree or under a shed, from which they would emerge barking at full strength, when a stranger appeared, and it was usually a case of the bark being worse than the bite. In winter they would want to share the house and fire with the family. If allowed to spend their time inside, they would take possession of the best spot in front of the fire or on the floor, and someone would fall over them at least once a night. A resentful yelp would indicate that the dog believed that particular spot was his by right.

A funny thing about dogs is that some seem to be natural hosts for fleas, but others appear to have no attraction whatever for these pests. A "fleabag" dog is not an asset, no matter which way you look at it. He might enjoy scratching himself, but no one enjoys the fleas he spreads around the place. In the old days there were few anti-flea powders and other pest-repellent substances and the dogs were let deal with the insects in the way they pleased.

Sometimes a dog would become mates with a horse and both would fret if they were separated. In most instances, however, they regarded all other four-legged animals as enemies, to be barked at whenever the occasion warranted.

With the average family, the dog had an easy and full life and lived to a ripe old age, to be mourned by all who knew him when he passed on.

The Land, 19 October 1961

Stealthy work of the sheep killer

Dogs can often be a nuisance and a danger, as well as useful units, on a farm. A good worker is always worth his weight in gold—if there is enough work for him. But if work is lacking, the dog is likely to go to pieces, to become savage, turn into an egg-eater and sheep killer.

Our Australian working dogs have been evolved for special jobs and are out of their element if they are not kept at work. This does not apply to the half-breed, mixed breed or that downright "mong" that used to be seen on most farms years ago. One of these dogs could be innocent and inoffensive during the day, good tempered and well mannered, a shining example of a good dog. But at night, he would sometimes slip away, possibly a couple of miles distant, when he would become a savage sheep killer just for the sake of slaughter. When engaged in this type of thing he would resemble a wild and hungry wolf and was likely to attack any human who attempted to interfere with him. Fortunately, such dogs were not numerous, but those that did exist caused great losses to the unfortunates who owned the sheep.

No one suspected the culprit at any time, until the owner of the sheep did some successful shooting one night and recognised the carcass as that of a neighbour's dog. Never at any time had the dog shown any inclination to interfere with the sheep on the home farm. You never knew what type of dog would turn out to be a sheep killer or when. Some would take to it early in life, but others would be well on in years when they made their first attack.

Two of the worst killers I ever knew were a Fox Terrier and a little mongrel of mixed breed, but with a fair share of Pomeranian. Operating a few years apart, these dogs must have killed hundreds of sheep before they were shot. No one would believe that the Pom type was a killer until, becoming careless, he attacked a mob of travelling sheep halted on the road during the heat of the day. The way he went about the job revealed long experience and the development of a technique that brought the best results in the shortest possible time. We often thought afterwards that if that small dog had been trained to work sheep, instead of training himself to kill them, he would have been the equal of any Kelpie or Border Collie.

Sometimes sheep killers would hunt in packs, causing terrific havoc for a short time, but they were never the danger the lone killer was. This was because a pack was easily seen, even at night, and they could be killed off with rifle or shotgun. They were inclined to run about and bark, instead of concentrating. Strangely, few pack killers would operate alone once the pack was broken up and, even if they did, they were too amateurish to have much success. However, the lone killer was hard to detect, stealthy and expert at his job and he had a disdain for baits laid in an attempt to poison him. To kill the killer was a battle of wits.

The Land, 26 October, 1961

"Foxie's" cunning life came to an early end

Taming wild animals is a specialised business and is not always successful. I know—I once tried to tame a fox.

It happened during my schooldays. We had a gang of men digging out rabbit warrens and they unearthed a vixen and one small pup. I was something of a favourite with those gentlemen, so they sent for me and on my arrival presented me with the pup. Their advice was to bring the youngster up as I would an ordinary pup, a job in which I had had recent experience. My methods proved successful, for young "Foxie," as we called him, thrived from the start.

One of my earliest trials came with kennel hygiene. Foxes have a strong, peculiar and penetrating odour at the best of times, but when kept near human habitation the reek from the kennel is really "on the nose." Natural odours are helped along by the fox's habit of cluttering the place up with incompletely picked bones, scraps of meat and so on. Housewives have a natural objection to this kind of thing and my mother was no exception. So a lot of my spare time was spent in maintaining the fox kennel in a state of cleanliness.

That fox was a cunning character. He worked in close cooperation with our old sheepdog. Their aim seemed to be to reduce the poultry population. Our fowls had open—wide open—range, and wandered all over the farm at will. The old sheepdog was a good worker, and he'd muster the fowls and drive them to where "Foxie" was tethered at his kennel. Then there would be a "snap" and another fowl became a casualty. Feathers in and around the kennel gave the game away, but it was not until a close watch had been set that the sheepdog's part became known. I was instructed to take strong protective measures if I wanted to keep "Foxie," so a netting fence was placed around the kennel.

"Foxie" had a soft spot for fowls and would snap at any strays when I took him, on the end of a stout chain, for his exercise walk.

Everyone seemed antagonistic about "Foxie" and described him as "useless," "a waste of time" and in any similar terms they could think of. In fact, they thought he could become a definite danger if he escaped, because his partial domestication would make him a particularly cunning enemy when the inevitable happened and he got amongst sheep or fowls. Those views weighed heavily with me when I accepted an invitation to spend the school holidays with my grandfather. My tentative suggestion that I take "Foxie" along got nowhere, so it was with a heavy heart that I parted from him.

As expected, I never saw "Foxie" again. When I returned home, I learned that he had broken his chain, burrowed under the netting and killed a couple of lambs in the bottom paddock. Dad had been lucky enough to have a gun when he went around the flock at dawn the next day and "Foxie" had fallen to the first shot. All that remained of my half-tamed pet was his skin, pegged out on the shed door.

The Land, 18 May 1961

War was declared on rabbit hordes

All this talk about domestic rabbits and their possible danger to rural industry calls to mind the attack we made on the wild variety of bunny in the Grabben Gullen district.

We'd taken over a mixed farm, mostly of that rich, chocolate soil that grows top quality potatoes, good oats, and, when given a chance, will grow pasture turning off fat lambs that rival the famous Canterbury variety.

But when we went there, the soil wasn't being given a chance, simply because it was overrun by rabbits. You'd find them in hollow logs, in the blackberries and in warrens that resembled what one hears of the Catacombs. Only the toughest tussocky grass survived—the rest of the place, if you forgot the lush growth of blackberries and briars, was as bare as the road. Even the bunnies were feeling the pinch: they were ring-barking the briar bushes by eating the bark from them and were attacking the young blackberry growth. About sundown the rabbits would begin to come on the flats and you'd see acres of their brown-grey backs and white tails, giving the impression that the earth itself was trembling.

Dad was not discouraged. Although only a small man, he had a big heart and he planned a campaign against the enemy hordes.

First job was to net the boundaries and for weeks wagons arrived with coils of wire and rolls of wire netting. The noise of the axe and the crosscut saw was loud in the land as trees were felled, accompanied by the sound of the maul on wedges as posts and droppers were split. Fencing was not an easy job as the boundary went up and down some rugged, rocky hills, but it was completed eventually and our share of the rabbits was enclosed on our own place.

Many of the neighbours thought we were mad spending so much money on a netting fence—they were content to have their boundaries fenced with long-legged timber and just to keep the stock in. Apparently, they were prepared to co-exist with bunny in his thousands.

Now began our real campaign, planned with the care of a military movement. As the burrows and warrens were ripped out, the smaller fallen timber was "picked up" and stacked against the logs. Soon the district was covered with clouds of smoke as our dead timber was burned. Horses were used to pull the partly burned logs together and the fires continued, with all of us looking like demons as we laboured among the flames and smoke until late at night. Brush hooks were called into action to slice into the blackberries, which were fired immediately the canes dried out.

During this period of hard toil, a motley pack of "rabbit dogs" was built up, including numerous breeds and crosses. As the harbor and homes of the rabbits were destroyed, the pack went into the attack, supported by shotguns and rifles that made crack shots of us.

As the place was cleared up, so were the rabbits, but the survivors became cunning and other measures were adopted. Poison trails were laid, but the greatest fun for us young folk was in the rabbit "drives." For these, a length of netting would be run out at an angle from the boundary fence and a team of "beaters," with yells and the aid of barking dogs, would drive the rabbits into the "V" formed by the netting. Another length closed the open end and then began the slaughter.

In time, only stray rabbits were left and the dogs soon ran these down. Of course, while a rabbit remained, a close watch had to be kept on burrows being scratched out in some of the soft soil and the fence had to be watched because stock often pawed holes in the netting. When that happened there'd be an invasion from outside and a lot of the work had to be done over again.

Some immediate return came from the campaign, as carcasses were sold to the nearest freezing works and, in winter, the skins returned a good price.

Those neighbours who shook their heads and said we were mad when our war began followed suit within a couple of years.

We had grass on our place, lush where the burrows and warrens had been dug out and where the timber had been burned. Our oat and barley crops were safe, our stock fattened and we had to go poaching on someone else's property when we wanted to have rabbit on the menu.

Soon it was a countryside of productive farms.

But if "domestic rabbits" escape and stand up to the conditions under which the grey-coated whitetails thrived and multiplied, there's more trouble ahead.

The Land, 9 March 1961

Rabbit trapping was a serious business

Rabbit trapping must be tackled seriously if it is to be a success; the habits of the bunnies must be studied.

It is easy to put out a line of traps, but not so easy to have them filled, although they might be set in country heavily populated by rabbits. A lot depends on the way in which the traps are set and how evidence of human intrusion is kept to a minimum. In my time, the expert trapper made good money—the others just existed. Lots of us did some trapping when we wanted to raise some extra cash or when the rabbits became too numerous on our own country.

A walk over the terrain would indicate where the rabbits were to be found, and where the traps should go. There is little point in recounting such details as they varied with the country, the season and the availability of feed.

The general plan was to set the traps in a line, with the far end curving inwards slightly in crescent shape. Traps needed to be in good condition so they would spring easily; otherwise bunny would escape. A shallow hole would be scraped out of the soil, generally on a track, near a squat, burrow or heap of droppings, all of which revealed that rabbits were about. These earthworks would be done with a "rabbiter's hoe", a short handled, narrow-bladed hoe, not unlike the entrenching tool used in World War 1.

The spike at the end of the chain would be hammered into the ground and the trap set—lightly to ensure that it would "spring" immediately weight was put on the tongue. A square of newspaper would be laid over the tongue and one jaw of the set trap to prevent dirt from dropping through, thus stopping the trap from going off. It was the habit of the trapper to cut sufficient squares of paper before he started on the line and to pin the pad on the front of his shirt for ease of operation. A thin layer of fine dirt would be sprinkled on the paper, sufficient to hide that part of the trap. Then the shank would be covered with some of the rougher dirt, bringing the level, if you were a good trapper, to that of the soil before you started setting. For an experienced trapper, all this took only a couple of minutes and he'd move on to the next site at a brisk walk with the traps, carried by the spikes, slung over his shoulder.

Anything from a dozen to a couple of hundred traps would make up a "line," but the more used, the harder the work. Setting would generally be done during mid-afternoon and there would be a "round" after dusk. This meant that the trapper went around his traps, inspecting each one. When a rabbit was caught, it would be extracted and placed in a chaff bag, still alive, if the trapping was for carcasses, to be killed, bled and gutted later. The trap would be re-set and the trapper would go on his way. Generally, there would be another "round" between 9pm and 11pm, when the catch would be heavier. About daybreak next morning, the final "round" would be made and it would generally be the most prolific of the day.

On this last "round" the traps would not be re-set—they would be sprung and gathered later on in the day for the next "line," which would be a chain or two in advance of the first one, working across the paddock. In theory, this spacing of the lines of traps drove the rabbits ahead and there was little doubling back when the trapping was handled correctly. After the final "round," the trapper would prepare his rabbits for market or skin them. But that is another story.

We hear of "rabbit farming" these days, so I suppose the trapper still finds a place for himself in some areas.

The Land, 6 July 1961

Rabbits kept our finances liquid

It was possible when I was a lad for a schoolboy or teenager to more than keep himself in clothes and pocket money as a rabbiter.

Necessary items of equipment were rabbit traps and a good pea rifle. Some expense was necessary to acquire the traps if none could be raked up on the farm, but rifles in those days were comparatively cheap and ammunition was not all that costly. Rifles were preferred to shotguns as a bullet did not do so much damage to the skin as pellets from a shotgun did. As a result of rifles being used to shoot rabbits, some good marksmen were developed, as the aim was to swing the weapon into action with the least possible delay and to kill with one shot.

On a hunting trip, rabbits would be found in squats. Their ears generally gave them away. Squats were found against logs, stumps, in tussocks of grass, near stones and, at odd times, right out in the open. The rifleman would get himself into range, raise the rifle, aim and fire, all in a matter of seconds. Some of us had a go at shooting at moving targets, but that method was frowned upon because of its dangers—you could hit a sheep, a cow, a horse, or even another human who might be in the same paddock. However, in clear country, such as on a flat late in the day when the rabbits would be found near a dam, shooting on the move was permitted and possible. It really taxed our skill because we had to be right on the target.

Senior members of the family generally preferred us to make trapping our main aim because it was not dangerous and really cleaned up the pests. We'd set a line of traps late in the day, go round them just after sunset removing the catch, make another round about 9pm and the final round soon after daybreak.

This was following the system employed by professional trappers, but was on a much smaller scale as our battery of traps was not so numerous. That 9pm round was not so good where I was concerned as I was scared of the dark and the ghostly glimmer of the hurricane lamp I carried did little more than create weird shadows that did not improve my morale.

When rabbits were removed from traps, they were usually placed alive in a chaff bag, carried over the shoulder. Traps were re-set. In good rabbit country it was possible to get a catch in the same trap on every round.

Taking rabbits home alive was done when the trapping was for the carcass; they were killed as soon as they were extracted from the traps if the aim was skins alone. Skinning the catch was a gory job. Anyone who has had any experience will know that rabbit flesh has a distinctive odour. When you are handling the bunnies every day for a period you can't get away from that smell, or get it away from you, no matter how much soap and hot water you use for washing. It permeates your clothes and the fastidious would always change into rabbiting rig for any of the jobs.

Skins were pegged out on frames—a U of fairly stout gauge wire—to dry, and there was some skill in this to retain weight, all of which told in the price received. That price was governed, too, by the grading you gave the skins. Different weights and sizes were put in different bundles. A separate lot was made up of any that were damaged. Skins were sold to a dealer in the nearest town, some of whom did big business, or were consigned to one of the wool firms in Sydney, where they were auctioned. For this, however, you needed a fairly good collection of fresh skins, but the cheques were always worthwhile.

Many a schoolboy has bought a suit, pair of shoes, overcoat, new rifle or other needed commodities from the proceeds of rabbit skin sales.

The Land, 28 June 1962

Money in pigs

Pigs—and the money that could be made from them—have always fascinated me. However, I have preferred to admire pigs from a distance, as it were, and fate has never turned me into a pig raiser.

I have seen money made and lost in pig raising and long ago came to the conclusion that it must be a highly specialised undertaking to yield the best profits. This was brought home to me when a friend of mine, who was developing a butchery business, decided that pig raising in a big way would be the answer to all his financial problems.

He acquired a stretch of flat, swampy and unattractive country covered with grass tussocks and low, unattractive shrubs of the ti-tree type, as well as a prolific growth of brittle-jack on the few rises. His view was that the pigs would help to clear the place and that their rooting around, plus the addition of pig manure, would give the soil fertility a great lift. In theory his ideas were good ones, but they took some working out.

He bought pigs—on credit—at auction sales over some weeks and liberated them on the new run.

His first worries came when the pigs found that the boundary fences were not in good condition, so a lot of work was spent on them. Repairs to the fencing proved so costly that ideas for subdivision had to be postponed indefinitely. The pig run was operated as one big paddock. It was not enough and, as the district maize crop failed that year, there were difficulties in providing a solid, more balanced ration.

Eventually my friend decided to walk out before he became too deeply involved. Just before the closure of the enterprise, I remember seeing him making his way through the shrubbery with squealing pigs racing away from him. He came over to the fence and admitted that he was beaten. "There's money in pigs," he said, "but it takes a helluva a lot to get it out of them."

Earlier, I had figured in some interesting experiences with pigs.

We always had a few pigs on the farm to provide bacon for the family and to add to our income, but we were never in it in a really big way.

On one occasion Dad came home from a sale with half a dozen black slips of the razorback type, a really unprepossessing lot. To allegations that they were wild pigs that someone had palmed off on him, he said they'd soon fill out when they got a bit of feed into them and they represented the opportunity of making a quick profit. However, he decided to keep the slips in the crate over night.

Released in the pig paddock, which had a high and secure fence around it, next morning, the newcomers occupied themselves by gorging on everything that looked like food. But they had all gone next day. They must have climbed the fence like monkeys. A ride around the farm failed to reveal any trace of them until we took a look at the boundary fence. There were six separate holes, at widely separated points, in the wire netting, indicating that the razorbacks had gone through like modern guided missiles. One had been seen by a neighbour, who complained about damage to his fences, but no one in the neighbourhood saw those pigs again. We did hear months later that a couple of wild pigs had been shot in the hills miles from our place.

It always pays to concentrate on the domestic type of pig in the fattening business.

The Land, 7 September 1961

A bonanza in bones

Collecting the bones of dead stock was one of the ways in which we built up our spending money in the years before prosperity really hit the country. It was not much of a job, in retrospect, but at the time there was little wrong with it as the return was good.

Normally there would not be many bones to collect—stock were too healthy—but a few could be picked up around the homestead from the sheep and cattle killed for home consumption. But in drought years, it was a business that boomed. Every paddock would be marked with a pile of bones where a sheep had died, and every one of those bones meant a financial return. Sometimes, but not often, cattle died and their skeletons gave us a good yield.

We struck a bonanza when we found, during a drought, that some of our neighbours had no interest in the bones of their dead sheep and planned to burn them, when they got around to it. We told them we'd save them the trouble and moved in to gather a harvest that was richer than the most optimistic of our team expected. It happened only once as the neighbours heard something about what we had made on the bones and decided to cash-in themselves in future.

In the actual collection of bones, we'd use a chaff bag if there were not many, carrying it across the front of the saddle. In bad years for stock, we'd drive around with the spring cart, tossing the bones in as we went.

We'd have a dump within easy distance of the house, but no too near because the smell from the heap of bones became "high" no matter how clean they had been picked. That smell, or stink, of bones was one to which I could never accustom myself. It seemed pervade everything. It was impossible to get rid of it from the hands and clothes, no matter how much soap and hot water were used in washing. Another disadvantage in the bone heap was the wide variety of insects that congregated and bred there. Amongst these were some evil looking weevils, whose sole cause for existence seemed to be eating the meagre scraps of dried flesh and gristle that might have remained on the bones.

When we had collected all the bones we could, we'd have a yarn with the local dealer and, after much haggling, a price would be set. He would arrange for the bones to be carted off and railed to the interests that bought them from him.

In really dry years, many truckloads of bones would leave our district, but the returns, plus money from dead wool, represented some saving, small as it might be, from the wreckage represented by the stricken stock.

One thing we learned was that bones had to be "green" to bring the best prices. They must be reasonably fresh and should not have spent too much time in the weather. With a big consignment you could get away with a few dried out bones that had spent years in the paddock, provided they were mixed in with the fresher ones.

After one of these consignments had reached its destination it was not uncommon for the dealer to make a few unpleasant remarks because he had to stand any loss. Not that it worried us much: we knew he was making a good profit, which is the reason why we eventually consigned bones direct to the city buyers, thus cutting out a middleman.

The Land, 8 March 1962

How we missed out on attaining big money

Do you ever pause to think of the big opportunities you may have overlooked in your early life, the opportunities, for instance, that could have led to fame and fortune?

This type of thinking is one in which I engage when I am feeling depressed, and it usually leaves me in worse shape than when I started. Most of my missed opportunities seemed to have occurred in my late 'teens and early twenties for some reason or other. Perhaps I have been too busy since then making a pound or two to think up any bright ideas, the reason why I have not joined the ranks of the nation's tycoons.

When I was a teenager there was the missed opportunity of the Big Bag Business, based on the purchase of second-hand bran, pollard and flour bags and their re-sale to neighbouring potato growers, at a profit, of course. In those days, no one seemed to care much how potatoes were marketed, just so long as the bags held together to get them from the point of dispatch to delivery. Being rather cautious, I made exhaustive inquiries about likely bag supplies and found that they would be adequate, for a start at any rate, and that a great deal of capital would not be required. In fact, the business could be financed from my rabbiting profits. However, my investigations meant too much delay and someone else hopped in, possibly given the idea by my activity. He did all right, too, eventually branching out into all sorts of associated business undertakings, including dealing in hides and skins, various agencies, and buying farms. To use an Irish term, he'd be a "warm" man today. But I must admit that he had always been a better businessman than I.

Sometime later I had the idea of developing a carrying business. That idea came when the farmers were giving up using their own teams to cart their potatoes and other produce to the railhead and were employing contractors. Behind my plans was the use of one or more motor lorries to do the job. I became involved in reams of figures representing running costs, capital costs, the charge I would make to the clients, and so on. Then a big decision had to be made about the type of vehicle to be employed—two-wheel drive or four-wheel drive. There was, of course, a big variation in cost, but the four-wheel drive appeared to offer a greater degree of trouble-free operation in wet weather, when most of the work had to be done. In the midst of all these calculations I suffered a severe illness which put all thoughts of business out of my mind. When I recovered almost a year later, I found that once again someone had started out with the same object I had been fostering. He went broke after a year or two—the farmers started to buy their own motor lorries and the people who made the money were the agents for the vehicles concerned.

At various times I had considered starting up a fruit and vegetable run to serve the district for, strangely enough, few of the farmers in our neighbourhood had gardens or tried to grow fruit trees. Again came the inquiries about supplies, a vehicle to do the run and the size of the territory it would be possible to cover. During the planning phase I remembered that a distant neighbour had had a go at a similar kind of thing some years before and I thought that a yarn with him would be helpful. It was. He advised against it, and reckoned I would not even cover my costs.

"But they all eat fruit and vegetables and they buy most of them in town," I protested.

"Sure they do," he replied, "and that's the snag you or anyone else is up against—the buying is an excuse for a weekly trip to town. They run accounts with the greengrocers there, so you wouldn't have a chance."

And another business failed to see the light, but I saw others go broke in attempts to develop it.

The Land, 26 July, 1962

A million doesn't mean so much these days

Do you remember the days when a million meant a lot—of money, people or anything else? Nowadays millions are common in practically all fields of endeavour.

When I started school the population of Australia was between four million and five million people and when the six million mark was passed the country seemed to have reached a milestone. Now we have more than 10 million, a figure that definitely is not big enough, and not nearly as awe-inspiring as five million was.

It is in finance that millions have become common currency. Once upon a time the companies that appeared to overshadow the whole financial structure of the nation were CSR and BHP, with a couple of others close behind them. Then came the Bank of NSW [now Westpac] and possibly one or two other banks, but no one seemed to know the actual capital of any of them. It would probably have staggered a lot of people to have known that any one of those concerns could quote millions when discussing capital and/or turnover.

Always on the horizon were the great pastoral firms that sold our wool and stock, sometimes loaned us money and served us in other ways. They, too, were suspected of having access to a lot of money, but not to millions.

Came World War II and its aftermath of expansion that made millions common talk in finance. Companies of several millions of pounds capital were floated, others lifted from a modest few hundred thousand to a million or more. All this was breathtaking to the conservative types who had been used to talking in thousands of pounds. More was to come with the take-over era, when millions began to look like small change.

I suppose these things happen in a rapidly expanding country. It is, however, a development with sobering effects, especially when one reads of wool sales over three days bringing a financial return of a million and a half in the old home town when one recollects returns from the earliest sales totalling a couple of hundred pounds.

Even with millions being a common term in most things, it is still difficult to visualise what a million looks like, whether in people, pound notes, pigs, or loaves of bread. Has anyone ever seen a million of anything in one go? It would be an experience because a million is still a lot after all.

We were discussing this subject the other day and someone said that it should be possible to make a million, provided the right idea could be made, given ability, a lot of luck and able assistants to cope with taxation problems. Yes, it is possible to make a million or to own a million of something or other. And it is possible to go broke for a million—another sobering thought.

The Land, 1 March 1962

A knotty problem

There are not so many reapers and binders about these days, but it is not so long ago that they were essential equipment on any farm producing grain and hay. Those were the days when cereal crops were cut with the reaper and binder and threshed or cut into chaff by plants owned by contractors. In districts where farms were big, the reapers and binders themselves were often operated by contractors, especially as in some types of weather it was essential to get the harvest in as soon as possible.

Haymaking was then a big industry—and it had to be, for the many horses used for haulage had to be fed and their owners demanded high quality oaten chaff. But the coming of the motor truck has changed all that—the area of oats has fallen and the demand for chaff has slackened off to a trickle, just enough to feed race horses and show mounts, as well as the few working horses that are kept on some farms.

And another reason for the decline in the use of the reaper and harvester is the rise of the combine harvester, which came in under the names of the stripper (nothing to do with the Gypsy Rose Lee and her ilk), header harvester and so on. Eyebrows would be raised today if any farmer started out with a reaper and binder to cut a wheat crop and then looked around for a thresher to complete the job. But it wasn't so strange 40 years ago, or even less, when the "binder", to give it the more popular name, was brought out every harvest and had to work hard for some weeks.

A careful farmer would keep his binder under cover from harvest to harvest and would find it in good order and condition when he wanted to use it again. Others, not so careful, would park the binder under a tree, where it would become a roost for the fowls and something that horses and cows could use for scratching themselves. These farmers would have to remove a lot of poultry manure from various parts of the machine before they could get it into action. Leather belts and canvases, if left on, would have to be replaced, and gallons of kerosene and oil used to clean up the rust and get the machinery running reasonably smoothly.

Like a lot of other machinery, timing is important with a binder, especially where the knotter is concerned. In fact, the whole of the machine seemed to centre around the knotter—the knives that cut the crop, the canvases that carried it to the section that knocked it into sheaves and passed the twine around each sheaf. All had to be timed perfectly with the intricate bit of mechanism that tied the knot and cut the twine. I have seen strong men raise their arms to heaven and almost weep when they have had everything else running smoothly, but the knotter failed to work.

It must be remembered that every farmer had to be his own mechanic where such things were concerned. It might have taken days to get an expert from town because he'd be busy starting a new plant elsewhere and, in the meantime, a good crop was becoming over-ripe. So the game of trial and error would go on and the language would beat anything that bullock drivers were reputed to use, even in their worst moments. Eventually success would be achieved, everything, especially the knotter, would be working smoothly, the sheaves flying off the board and the stooks rising rapidly. Of course, the driver might forget to oil the knotter and to clean any odd straws away from it, which meant more trouble.

The Land, 24 May 1962

Chaff cutting, threshing brought busy farm days

Progress in the form of mechanisation has removed from the Australian rural scene two once familiar sights—the itinerant chaffcutting and threshing plants.

Chaffcutting on a major scale is no longer necessary because of the decline of the horse as a power unit in the farming and transport industries. And it is far more economical to thresh grain in the paddocks with the modern header than it was with a threshing machine.

Both the chaffcutter and the thresher were powered by portable or traction steam engines. Where a portable engine was used, it was drawn from farm to farm by a team of horses, as was the chaffcutter or threshing machine. When a traction engine was used, it drew the other unit.

Those traction engines were impressive sights, with their large iron rear wheels to which heavy steel treads were fitted. Smoke belched from their chimneys, there was a hiss of escaping steam and they trundled with a lot of rumbling over some of those stony and rough roads we had then. However, the driver of the traction engine was the envy of small boys for miles around, any one of whom would have given all he possessed (and it wasn't much) to be up there in the cab at the controls.

Chaffcutting or threshing time was always an event on the farm. Bags and twine would be on hand, a load of wood tipped near the stacks and a water cart ready for the job. In the house, Mum and the girls would be cooking gargantuan meals and preparing to brew many gallons of tea. Arrangements would have been made with the neighbours for the working gang, none of whom was paid. All would be farmers "giving" a day or more to the job, their repayment being the equivalent time from the man they helped—the true co-operation of country people. The plant was owned by a contractor, frequently a farmer himself, who, with a couple of assistants, had to be paid.

Came the big day. The plant would be located in a position of vantage among the stacks, which had been built in a bunch to avoid carting the sheaves or long pitching. Care was needed in lining up the engine and the cutter or thresher because the line of the driving belt between both had to be perfectly straight. Otherwise, the belt would naturally fly off.

Members of the working team would take up their posts. One of the contractor's permanent staff would drive the engine and another would feed the sheaves into the machine, cutting the bands of binder twine as he did so. This was a skilled job, as speed was needed as well as care with the razor-sharp cutting knife. He'd hold each of the bands in one hand until he had a fair bundle, when they would be thrown to the ground, to be retrieved later for various tying jobs around the place. Several of the gang would pitch sheaves from the outside of the stack to a fast and accurate pitcher, who, with a short-handled pitchfork, would drop each sheaf on to the feed board in the right position and at the right time for the feeder.

Behind the machine would be a bagger, whose job it was to fix the empty bags on to the hooks at the outlet of the grain and chaff hoppers. As the material poured into a bag, it was necessary to shake it to ensure that the bags were well filled. When filled, the material would be diverted to the other outlet, the filled bag pulled off and swung around to the sewer. With a flying needle, he would stitch up the top of the bag and dump it on the growing stack behind him. From there it would generally be carted by dray or wagon to the shed or barn, (to use the correct but neglected title) for storage.

There was another job to be done at threshing time—the straw had to be put into some kind of orderly stack as it came from the elevator after the grain had been threshed out.

All the time the engine driver had to keep his machine stoked up to maintain a good pressure of steam and to make sure that the cask from which the engine drew its water was full. Water was drawn in a couple of other casks, or a square iron tank, on a dray from a dam, spring or tank, a job that was the responsibility of the farm's owner.

Came "smoke-oh", or mealtime, and the engine whistle would be blown. All hands would stop work and attack the food and tea prepared by the womenfolk. There would be a few minutes for a smoke and the whistle would screech again for a resumption of work.

On a big farm, in a good season when crop yields were abundant, chaffcutting and threshing could go on for several days, but on most places in my old district a couple of days would see the job out.

You don't see the thousands of tons of hay made for chaff these days, as the market has vanished. When the horse provided the power, the farm teams had to be fed and they needed some chaff to keep them in hard condition. But the bulk of it went off the farm to be sold in the towns and cities where horses drew everything from milk carts to wool wagons.

Grain is now stripped in the paddock and one header will do as much in a day as the old time method of cutting the crop with a reaper and binder, carting it in and stacking it, then threshing the sheaves, could accomplish in a week. In a few words, the eclipse of the horse has meant that cereal production is streamlined.

The Land, 6 April 1961

Country saddlers making a comeback

Among the skilled crafts that had almost vanished from country districts a few years ago was that of the saddler.

It seems that the mechanised age, which had replaced the horse as a working unit, had dealt a death blow to the saddler. That was true in many instances, as you will not find a saddler in numerous centres, but in others the craftsmen have held on and are now making a comeback. This is because of the rise in popularity of the horse for leisure time recreation, fostered mainly by Pony Clubs, and the increasing interest in trotting as a sport.

Horse numbers are increasing everywhere, as a visit to any show, from the one-day bush fixture to the "Royals" of the capitals will reveal. And the rapid rise in the breeding and racing of trotters and pacers is another big factor in the increasing population of light horses. Success of Australian pacers in the United States will focus still more attention on the light harness sport and give it the publicity that will bring further expansion.

We have always had big stables of thoroughbreds in the capital cities to cater for the followers of racing, but many country racing clubs have been battling for years. Picnic race meetings are becoming more popular and are gaining support throughout the country—another indication that the day of the horse is by no means done.

People who own and use horses for recreation and sport generally demand gear of a higher quality than that used in the more rugged days, when horses were used for farm and station work, droving and similar jobs, as well as for sport. This means that today's saddler caters for "high fashion" ideas in saddlery and there is a trend towards hand-made gear.

Some of the old-time saddlers have come out of their temporary retirement and are training youngsters in their skills so that the craft of the saddler will be a continuing one.

As in all things, the individual can add distinctive style to hand-made saddlery that is lacking in the mass production article. Because of this, one learns with more than passing interest that saddlers, working as the old-timers did, are producing completely hand-made and hand-sewn saddles, bridles and other gear. Other products for the horseman that have not always come from the saddler's shop have been whips, plaited reins and belts. Some saddlers went in for this kind of thing, more or less as a sideline, but it was usually the job of a specialist. Around country saleyards and showgrounds, you'd see a chap with a collection of plaited work and, to attract attention, he'd crack a stockwhip. At one time his display drew as much interest as the latest model cars do these days. You could buy a whip, a pair of reins or a plaited belt from him on the spot, but if you were really fussy you'd place an order for the job to be done in accordance with your own design and your own ideas of materials. Many men working on farms and stations, as well as drovers, went in for plaiting as a spare-time occupation and turned out some excellent work for themselves and their mates. Hours of practice were necessary to become adept at plaiting the fine strips of leather, but the experts could produce faultless work. Many of them are still doing it.

All this leather work, from saddles to fly-veils, is in demand today and the demand will grow. We can't see plastics being used for saddlery and harness— the horse-lover would miss the tang of the smell of leather blended with sweat, the smack of the straps and the creak of the saddle in a strenuous horse event.

The Land, 1 June 1961

Saving the horse from oblivion

A few years ago, it seemed that young Australians would never know how to care for a horse, and would never, in fact, know much about horses. All this despite the fact that the horse has played an important role in the opening up and development of the country, to say nothing of defence.

Horses were becoming fewer and fewer on farms and stations and it seemed that the final blow had been struck when we heard of motor bikes being used for boundary riding and mustering. To the old hands, this meant that the final strongholds of their equine friends had been breached.

About that time, however, the Pony Clubs began their spectacular rise in popularity.

All over the country, young people began to take an interest in the horse for recreation and pleasure, finding that competition in show events added some point to their ownership of a pony or hack. They found, too, that membership of a Pony Club taught them a lot about the care of horses and about show competition.

The Pony Club movement began to spread from district to district like a bushfire, until today we find that the real backbone of the ring section at practically all country shows is provided by Pony Club members. It does not stop there, either, because the Pony Club people grow up and as they do, they retain their love of the horse and continue their support of shows, not just to win ribbons but for that feeling of attainment that brings progress to any enterprise.

In their club and show work, the young people are learning things that became second nature to many of us who were born and reared amongst horses, who could ride almost before we could walk, and who could weave our way as toddlers between the legs of the family's favourite steed, or even under the bellies of the farm team. They are learning the importance of correct grooming to the horse's welfare and are able to wield curry combs and brushes in expert fashion. They are learning that a horse must be cooled off after vigorous exercise and that a roll in a sandpit or a heap of dirt helps in the cooling off and contributes to the general health of the horse.

After the sweat has dried comes the hard work as the curry comb removes the worst of it, generally to the accompaniment of a hiss, which simply means that breath is being expelled to prevent the dried sweat from being breathed into the lungs. Then comes the brush work with a stiff brush to remove the last vestiges of sweat, followed by the soft brush to bring up the shine on the coat. For the really fussy groom, the final stages will be with a soft cloth, preferably silk, which adds a gloss to the coat.

Despite the work involved, grooming a horse is a satisfying sort of job for the horse-lover. He knows that his charge is enjoying the experience and he gets a lot of satisfaction from putting the horse's coat into such good condition.

A well-groomed horse has a shining coat that reflects the variations of the sunlight as he moves and sparkles in the dark with static electricity as a hand is rubbed over it. Good grooming removes scurf from the horse's coat, down to the skin, but a horse that is incorrectly groomed will carry dirt through the hairs and on the skin.

Grooming, taken in one's stride as part of the job when working a team of horses or caring for a hack, is now a pleasant task for the younger generation of Australian horsemen and women. It is good to see them fussing over their horses because it shows that a love for a grand creature still exists in these days of high-powered cars and mechanised farm plant.

The Land, 28 September 1961

Bareback riding was sometimes a menace

Riding bareback is all right if you are used to it, but if you come at it green, you'll suffer.

When I was a lad, I did a lot of bareback riding, which meant that my seat had become hardened and my general physical structure was accustomed to that form of the equestrian art.

To ride bareback saved a lot of time on the farm—it was necessary only to put a bridle on the pony, jump on and round up the milkers, plough horses, sheep or go for the mail. For such short excursions, bareback riding was fun, especially as the pony seemed to enjoy the experience. And, as one of my younger brothers said, you did not seem to have so far to fall when you rode bareback as you did with a saddle.

Some chaps I knew were able to ride without either saddle or bridle. They had their ponies so well trained that they would obey the pressure of a knee as quickly as other horses responded to a gentle touch on the bridle. With those chaps, the no saddle and no bridle business became a craze and they endeavoured to train all their horses, including the farm work team, to do it. They did reasonably well, too, although they wasted a lot of time. But they baulked when it came to the evil-tempered trotting stallion they could hardly ride with a curb bit and a good saddle.

Bareback riding was at its best when the pony walked or cantered, but the in-between gait of a trot was purgatory. It shook you up and you seemed to wander all over the pony's back, as it was impossible to keep control with the reins—and the knee control just went by the board. If you got a fractious mount, you didn't have a chance of stopping on him for any length of time, even if you were a skilled roughrider. It is amazing what comfort a good saddle can be under such circumstances, and how much an aid to keeping your balance you find in even a surcingle. But, like all methods of riding and all forms of gear, bareback riding is a subject for never-ending argument and someone always knows a chap who tamed the toughest outlaw west of the ranges simply by his bareback riding skill.

My most memorable ride bareback was made, without premeditation, when I was about 11 years old.

It was a hot, humid summer night and I had ridden to a picture show some 12 miles from home with a couple of older fellows. My pony apparently got tired of being tied in the usual stall behind the hotel and, somehow or other, untied the bridle rein. He wasn't there when the show ended, and we knew him well enough to realise that he had gone home early—we later found him grazing in the cow paddock near the house, with all the gear undamaged. Anyhow, my only way of getting home was by doubling up, that is, riding bareback behind the saddle of another rider.

As I have mentioned, it was a hot and humid summer night and it was not long before I began to develop a prolific crop of sweat blisters on a section of my anatomy, and it was not long before they began to burst. The pain was memorable, but I preferred to put up with it rather than walk the remainder of the distance. Although the moon was shining, it was not a night when I could enjoy the beauties of nature. Next day, I was in a parlous condition. The sweat blisters were still, as they say, "giving me curry," and I was as stiff as a board. Since then, I have never believed the stories of the suppleness of the pre-teenager, able to take a lot that adults can't, because it was weeks before my physical condition returned to anything like normal. And I have kept bareback riding to a minimum since that night.

The Land, 15 March 1961

Cleaning harness was a wet weather task

Cleaning the harness was generally a wet weather job when we were running a horse powered farm, but it was not too pleasant if the weather happened to be cool. My memory indicates that most harness cleaning days were cool to cold and this took the temperature from the water we used.

First step was to wash the harness, all of it, with soft soap, which came in fairly large tins, and was applied with warm water. All pieces of the harness were unbuckled and washed thoroughly then wiped over with a cloth and hung up in the harness room to dry. It was during the washing that the water would cool off, but Dad never favoured throwing it out for a new lot too soon. He contended that this would mean an unwarranted waste of suds and, therefore, of soft soap, but he did agree to topping up the buckets and dishes we used with hot water. This kept things a bit more agreeable for a time, but did not last long.

Once the leather had dried off, the next job was to oil the harness with a home-made mixture. I have forgotten the recipe but I do remember that the main ingredient was neatsfoot oil. This commodity is renowned for its ability to preserve leather and it is renowned, too, for its remarkable and pervasive odour, which did not make the job a pleasant one.

It was necessary to rub the mixture well into the leather so that it would retain the suppleness that the soft soap gave it. Now rubbing an oily mixture into dozens of small pieces of leather is a monotonous job in anyone's language and conversation soon gave way to an intense desire to get the work over. Special care had to be taken with the riding gear, of which we were vastly proud. Saddles were reasonably easy to cope with because they offered a fair amount of surface on which to work, but bridles, breastplates, cruppers and other appurtenances were made of numerous pieces of thin leather straps, all of which had to get the best of treatment.

Somewhere through the job all bits, buckles and other silvered or brass pieces of the harness had to be polished until they glittered and all signs of the polish wiped away from the adjacent leather. This took time and there were some awkward pieces of harness for the working horses, particularly the wagon team, that made heavier demands on our time.

On harness cleaning days, the lunch break was not a long one, as Dad was always anxious to get back to the job and get it cleared up. Once clean, the harness had to be put together, which was not difficult if the pieces had been placed carefully at the beginning of the job, but it was somewhat laborious, coming on top of the day's heavy work.

On a good day, we could complete the whole job of cleaning, but if anyone felt a bit off colour, some would have to be left until the next wet day when it might only need and hour or two to go through.

One compensating factor was that our harness always looked well and lacked those splits and cracks showing up in the gear owned by less careful neighbours. Cleaning regularly kept costs of repairs and replacements down, but such points do not appeal to young fellows, as I was then, who regard harness cleaning as an unpleasant task.

The Land, 21 June 1962

When trotters were cart horses, too

Some of the pampered trotters and pacers of today would get a shock if they were subjected to the treatment that their predecessors received in the earlier days of the light harness sport.

Many a good horse has gone straight from a milk cart or a baker's cart to the course and won a race with no further preparation than a rub over with a brush and cloth to get his coat in order. Of course, the work those horses did fitted them for trotting, especially as the roads were mainly dirt, which helped to avoid leg and fetlock troubles.

Most of the cart horses started their racing careers in the country, but some of them raced at Harold Park, straight from the cart. It was to the milkmen and the bakers that credit for keeping the sport going must be given, with some kudos for the butchers. They bred trotters as cart horses, regarding them as utility steeds, which they were. In those days, of course, most of the light harness racers were square-gaited trotters as the hopples had yet to come into general use, but some good natural pacers were produced.

Around country towns, you'd see delivery cart horses wearing rubber bell boots to protect them from injury and you'd know at once that here was a horse being readied for racing, either at one or other of the few local meetings or at a coming show. Being fed to work, they were fed to race and were given little other preparation until a week or so before the meeting, when they would be given track experience. Most of them were ridden, as the racing sulky was regarded as an expensive luxury. But that view changed when most of the winners turned up in gigs. Those mixed fields of ridden and driven horses would horrify the star reinsmen of today, but we took them in our stride and there was some hostility when decisions were made to confine some races to the saddle and others to harness.

It was surprising where you would find trotters, apart from the delivery carts. Many a good performer got his start as a stock horse. Or as a general farm hack. In one district, the outstanding horses came from an orchard. They were well bred and looked the part, but most of their training was in front of the plough. They proved ideal for drawing the light orchard plough, which toughened them up and developed their stamina. A track on the orchard was used to school them for racing and they rarely had any experience in strange company until they were given their first start.

Distant meetings were no problem. The horse would be harnessed up in the sulky, which was laden with gear, and we'd set off to drive anything up to a hundred or so miles, probably leading another couple of starters, which would have their turn between the shafts before the trip was over. Nowadays, the horses travel in special padded trailers hooked on behind the car, by train or by aircraft. It would certainly be difficult to swim a starter to New York, but times and the outlook change and today's trotter has become a racing machine rather than a utility horse.

The Land, 20 July 1961

Reputation was not justified

It was by sheer accident that I attained a reputation as a horse-breaker during my teens.

Until the big day, my only hacks had been quiet stock ponies and similar mounts, which, when fresh might pig-root or engage in a few playful bucks. There was no viciousness about them and it was a friendly contest between us. Sometimes they tipped me from the saddle if they caught me unaware, but more often than not I stayed on them.

However, the morning I rode the Merv filly everything was serious, especially that little smokey grey. She had supposedly been broken-in by a neighbour, but she had a wicked eye and I never liked the look of her.

On this particular morning—it was cold and frosty, I recall—an urgent job had to be done. We had kept the Merv filly in the yard overnight and she was probably feeling as cold as I was, but Dad and I did not think of that as we saddled her up. She stood quietly but I did not like the way she trembled when the saddle hit her back. I was especially careful about tightening the girth and surcingle and I made sure that the stirrups were as they should be. In truth, I was fighting for time because I did not like the thought of mounting that filly in cold blood, but Dad encouraged me and held her head as I swung myself into the saddle. Somehow or other my foot found the stirrup on the off-side just as she started to buck. Dad hung on, but I had a fleeting glimpse of a piece of skin being stripped from the back of his hand and I yelled, "Let her go!"

When she found her head and was free from the additional restraint, the filly bucked in earnest, pulling downwards on the bridle. For the next few minutes she bucked solidly in the one place and I used all my strength and everything I knew to beat her. It lasted only a couple of minutes, but it seemed half a lifetime. Then she got rid of me—pelted me over the six foot fence and bucked herself to a standstill. I landed on my back with the wind knocked out of me, but my blood was up. I crawled through the rails, grabbing a stick as I went, caught the filly and was in the saddle again before she knew what had happened, But she did not make even a half-hearted attempt to buck—the spirit had left her. I yelled to Dad to open the gate and, as it swung clear, I sent the filly through like a bullet. I rode her hard for a couple of miles, which tamed her properly. However, she always had that streak of viciousness and was likely to put on a turn any morning so we got rid of her.

Somehow or other the neighbours got to know of my exploit and wanted me to break-in their horses. Pride would not let me refuse, but I approached every job in a more-or-less terrified mental state, which probably brought me more falls than I should have had. Time eased that a bit but no one was happier than when circumstances called me to another district where my reputation was unknown and I was just a capable horseman. I had never had any ambition to be a buckjump rider.

The Land, 31 August 1961

From the rough bush track to the tar-sealed highway

One of Australia's success stories can be found in the development and improvement of roads, particularly the highways and feeder routes.

This is not to say that our roads generally, and some sections of our highways, are beyond reproach. But in a country with a comparatively small population and in a period marked by two world wars and a major depression, the progress in road building over the past half century and more has been tremendous. Most of this development has followed the advent of the motor car and the improvement to motor vehicles generally. Within the memory of many people, most of our roads were little better than bullock tracks, kept in trim by shire workers with horse and dray, pick and shovel.

For decades, the early roads were little better than the first route inland, constructed from Sydney to Bathurst over the Blue Mountains by William Cox and his convict workers. But as settlement intensified and more stage coaches used the roads, improvements were made to provide swifter, smoother travel. It was found, too, that the bullock and horse teams needed better roads as they had a tendency to bog in wet weather. So more gravel was used to ensure a firmer surface and to reduce maintenance.

Then came the use of blue metal for macadamised roads, which certainly provided a solid surface for heavily shod horses, but they were not too popular with the bullockies, whose teams developed sore feet. Lighter traffic and saddle horses found the going hard on the blue metal roads in some areas and this led to the evolution of side tracks. These tracks, in the soft ground, ran beside the road proper, and were ideal for unshod horses and for lighter steeds that were shod. Of course, heavy rain meant gullying of the side tracks, which at times became so rough that a new track would be brought into existence. With the development of motor transport, however, the use of the side tracks ceased as fewer horses were seen on the roads.

Roads in those early days of motoring were nightmares. The "horseless carriages" must have possessed great stamina to keep going on them, literally bouncing from pothole to pothole and from rut to rut. Even in the years between the wars, motoring was not all joy on many stretches of roads, especially those that developed the dreaded "corrugations," which as many drivers said, "shook the guts out of the car and its passengers, too."

However, our road builders had the answer. Smoother, bitumenised surfaces came but much of the earlier work in that direction was wasted as the new surface did not bond well with the solid macadam foundation.

In time, roads were entirely reconstructed, being topped with a bitumen seal, which facilitates repairs and maintenance. Some concrete roadways have been laid in Australia, but they have not been generally popular. They last well for long periods, but are not easy to repair when consistently heavy traffic forces that need. Possibly, modern methods with concrete could bring improvements. In these days, all over the country, gangs are busily engaged in making still further improvements to the roads that started as saddle tracks for early settlers. The roads have passed through the stages of gravel, macadam, tar surface, until today they have reached the new and apparently successful "hot mix" bitumenised surface being used for many highways.

When travel was slow, it did not matter much whether a road was straight or wound its leisurely way around curves and ridges and between the boundaries of properties. Now, however, we have attained the age of speed which demands that we get from point to point in the straightest possible line. That is why we see huge reconstruction works and realignment of roads to take out many of the bends and to provide easier grades. And the work will continue as the road builders push out further into the interior to provide the all-weather routes that future national progress demands.

As with all other improvements, road building is a matter of money. Given adequate funds, we will have more and better roads, but we must be prepared to pay for them.

Truly, the bringing of our roads to their present state is one of the great stories of the progress and development of a young country.

The Land, 4 May 1961

When every macadam road had a side track

Do you remember the tracks that used to run beside the macadam roads in the days of the horse? They were soft earth, designed to save light horses from leg and fetlock trouble and were used by people riding saddle horses or driving sulkies or buggies. Teams of heavy horses used the macadam. They had to because the loads they pulled needed a hard surface beneath them or they would bog in wet weather.

Every macadam road had its side track, which sometimes followed a route yards away from the road itself. On some stretches, there would be a track each side of the road itself. I have an idea that the tracks came into being because some of teamsters rode saddle horses and would naturally be off the road as they watched their teams. Be that as it may, the tracks were used extensively by the lighter horses and it was pleasant to jog along one, especially behind a smartly stepping sulky horse.

Shire Councils, apparently, did not regard keeping the side tracks in order as part of their duties. This meant that the track at times got out of repair, so another would be exploited nearby.

As the tracks encouraged a certain amount of soil erosion, they would develop into somewhat deep gullies after heavy rain. This could be overcome by running the wheels of the sulky slightly to one side of the original route. It would not be long before another track was in being. Then the badly washed out tracks would gradually fill up, or become worse, depending on the nature of the soil, until they had to be abandoned altogether.

When it became necessary, through difficulties of the terrain, to ride or drive light horses on the macadam, the good horseman always stood out. He would take his charges carefully along that stretch, prepared to waste time and to make it up when it was possible to get on the soft track again. There was some thrill, however, to the youngsters to be riding or driving at night behind another horseman whose steed would strike sparks from the blue metal of the macadam as its shoes hit the surface. To go clattering along the hard road late at night was one of the favourite dodges of the young daredevils of those days, the predecessors of the motorbike riders of a later age. However, this kind of thing had serious effects on the condition of the horses whose legs soon "bunged-up" if they had much of it.

When some of the locals went in for trotters, the side tracks were ideal for road work, done in saddle or sulky because the now generally used gig had not been widely adopted. Around picnic race time, it was not uncommon to see so-called gallopers being worked in the same way.

With the motor age, the side tracks have gone and you can travel thousands of miles today without seeing even the traces of tracks so well-used only a few years ago.

The Land, 4 January 1962

Worked like a machine as he dug potato crop

Inspection of a potato digging machine at the last Royal Show recalled some of the methods used before mechanisation came to the spud industry.

This machine does everything—digging, sorting and bagging—and the crew rides at work.

Vastly different were the old methods, still followed by some, of digging with forks, picking the tubers into buckets, grading them as you went along, bagging them as the buckets filled. It was hard, backbreaking work, profitable only to the strong, experienced digger. I have vivid memories of one such digger, who handled our crop every year.

His name was Harry, he was of German descent and he lived in Surry Hills. A tall man, he was marked by a prominent chin and a black, Kaiser-style moustache which turned up at the ends in true Prussian style. Spare of frame, he was possessed of great strength. What he did in the off-season we did not know—all we were interested in was that he was easily the best potato digger operating in our district. Harry was a fast worker, seldom speared a tuber with a fork prong and graded the potatoes with comparative care. And he could dig and pick 20 bags a day, which was good going. He was possibly one of the first margin men in primary industry, for he was paid one shilling and threepence a bag when the general rate was a shilling, but he was worth it.

Harry used a long-handled fork, upon which he lavished care comparable to that given his rifle by a crack marksman. He would be out of bed before dawn, waiting in the paddock for sufficient light to see the dead tops of the plants so he could start digging. When he started, he would use that fork as if he and it were part of a smooth running machine. When the time came, after several rows had been dug, to pick up the potatoes, the same poetry of motion would be apparent. With a bucket on each side, he would go down those rows of exposed potatoes, picking the large ones into one bucket and the smaller grade into the other, with never a mistake. Unlike many people on farms, he disdained yarning as he worked—it was a deadly serious business for six days a week, from dawn to dusk. Harry rested on Sunday, if you excepted his weekly washing and the cooking he did for the following week.

He was an excellent cook, too, and would have made big money in shearing sheds. His aim was to cook meals that could be warmed up through the week to save time and his camp oven ran hot all day Sunday. The amount of food he cooked and consumed was amazing. He had little trouble in preserving it, for the winters, when potatoes were dug, were bitterly cold. When Harry pitched camp, he ensured that his tent was well protected from the weather and he always had a galley with a big fireplace. Food, he explained, was the most important factor in the life of a man who wanted to do a good job in the hard manual task of getting potatoes from the soil to the bags.

But eating and working were not his only occupations—he was a voracious reader and a keen student of world affairs. He liked nothing better than talks and arguments during evenings and on Sundays as he washed and cooked. If there was no one to talk to, he read and one of the difficulties was to keep him supplied with literature.

These new machines may do the job in smarter time, and possibly more efficiently, but they lack the personality of diggers like Harry. There are few men of Harry's stamp around these days, so it is just as well that the machines have come—otherwise we might have to do without potatoes.

The Land, 11 May 1961

When our potatoes were pit stored

A short and interesting film on farming in Denmark that I saw recently showed how farmers there stored beets in "clamps." Those clamps were the same as the "pits" in which we used to store potatoes after they were dug.

The word "clamp" is generally accepted overseas, particularly in Britain, and it is possible that "pit" was a local term. Be that as it may, our pits served a useful purpose because we were able to store potatoes in good condition for fairly long periods. This was important when the market slumped at digging time and looked like improving later on. We then stored most of our potatoes in pits and waited for a rise in the market, which often went up as much as 10 pounds in a few weeks. Another value of pitting was seed potatoes, pending grading, which was not always possible at digging time.

The method was to tip the potatoes from the bag in a long row, four to five feet high, in the shape of an inverted V. Sometimes a pit would run for as much as 20 or more yards, depending on the size of the crop and the quantity to be held. It was surprising how the tubers would conform to the shape of the pit as they were tipped. We only needed to tidy up those on each side of the bottom. After the day's tipping had been done, the pit was covered with several inches of straw or grass. This was essential to protect the potatoes from frost and wind damage that would send them green and affect them in other ways.

Careful farmers, who covered the straw with earth, preferably sods, gave such protection only for the shortest of periods. When the pit was correctly earthed it would shed the heaviest rain, and keep out frost and wind.

As I remember, some of those pits would be growing a lush coating of grass before they were opened up. If the potatoes were in top condition when pitted, they would open up the same.

For successful pitting it was essential not to include tubers showing rot and to exclude any with fork damage if possible. Otherwise the rot would spread and bags of potatoes would be lost, cutting down the profit margin. It was, of course, necessary to pit only the old potatoes—those dug after the frost had killed the tops. New potatoes—those from which the skins can be rubbed—cannot be stored in pits or otherwise.

Sometimes it was possible to have the pit under shelter. The main advantage of this was apparent to those who had to pick them over and bag them. This operation was normally regarded as a "spare time job" meaning that there was no spare time on the type of farm that included potato growing as only one of several branches of operation. You'd use a folded potato bag as a knee pad and go to it, sorting and grading the potatoes as you picked them up. It was backbreaking work and played the devil with one's hands in cold weather, as the tubers were always icy cold. But the job had to be done. However, it was quicker work than picking up potatoes in the paddock because you only moved ahead as the "open face" of the pit was attacked.

If the potatoes were left pitted too long they would have started to grow when the pit was opened. This meant that shoots had to be rubbed from them as they were handled. These potatoes were marketable but when the tubers began to wrinkle and shrink their market value was lost, although they made good seed. Incorrect covering of the pit would leave a thick layer of greened tubers which had little, if any, market value but they, too, were useful as seed.

"Pits" or "clamps", the system was the same as that used by the Danish beet growers.

The Land, 17 August 1961

"Hoofbeats" that speeded rider on

At a social function recently, I met a chap from the Southern township of Rugby who would have made a great public relations man for a centre of much greater size. He was so keen on Rugby that you would have thought it was the place where that brand of football originated, rather than a smallish settlement in NSW.

With me was another chap from the South. He was a better talker than I am but neither of us could make any contribution at all to the discussion. Both of our families had been associated with the Southern districts for almost a century and a half. We could each have added some colour to the discussion, but we were not given a chance by our young friend.

He told us about the Rugby show, which he claimed was the best little one-day exhibition in the world, a statement with which we had no quarrel. He praised the sheep of the district and everything else it produced, from glamour girls to sub. clover. No one expressed any doubts about any of his statements— in fact, we all agreed with them, having seen Rugby and its products at some time or other during our lives. But there were things we would have liked to have said about the place, if we had been given the opportunity.

For instance, I would have liked to have told the story, handed down from my maternal grandfather, about the bushrangers who once frequented the Rugby and surrounding districts, but I didn't get a chance.

Grandfather used to tell the story of one bushranger, rather a villainous type of fellow, who did not like actually killing a man as well as robbing him.

There was the story of the butcher named Slocombe, who fell foul of the bushranger. Slocombe was out on a stock-buying trip and was well equipped with golden sovereigns in his saddlebags. Our bushranger friend made Slocombe dismount and then demanded to know where his money was. Slocombe refused to tell him, so, being a powerful type, he proceeded to tie the butcher to a tree. All this time Slocombe was thinking that if the bushranger was going to kill him, he would have done it early in the piece, so he had a good chance of getting away and keeping the sovereigns—if the bushranger did not take his horse. Slocombe continued to refuse to say where the money was, so the bushranger produced an ugly looking knife that, as well as being blunt, had several gaps in the blade, an insulting thing to Slocombe's trained butcher's eye. He proceeded to saw at Slocombe's throat with this piece of dejected cutlery. Fearing blood poisoning, if nothing worse, Slocombe told him where the sovereigns were. Dropping the knife, the bushranger opened the saddlebag, removed the coins, mounted his own horse and cleared out, leaving Slocombe tied up for some hours until a rare traveller came along the road to Rugby.

That experience, told to him by Slocombe, made an impact on my grandfather, who was always scared when in the Rugby district, especially at night.

One night, however, he was jogging along when he heard what he thought were hoofbeats behind him and blamed the bushranger. He spurred his horse on and the pursuer did the same. This went on for some miles, and grandfather began to get a bit anxious, especially as the night was too dark to see anyone coming behind. If he steadied his horse, the pursuit steadied too. The game of cat and mouse went on for some miles. Eventually, grandfather, after rousing a canter when the sound of hoofbeats behind him sent his heart beating faster as he thought of Slocombe's experience, stopped his horse. His "pursuer" stopped, too. And the same time, the long, knotted ends of the neckerchief grandfather was wearing dropped to his back and then he remembered that they made a noise like hoofbeats when he was riding into the wind. No bushranger had been following him, and there was no reason for him to be scared for the goodly collection of sovereigns in his saddlebag as he rode away from Rugby township.

The Land, 17 May 1962

Search for Ben Hall's treasure ranged wide

When treasure trove is mentioned, it conjures up visions of pirate hauls that have been hidden for centuries, defying all modern attempts to find them. Many people have spent thousands on unsuccessful attempts to recover galleons full of Spanish coin, gold bars, precious stones and other treasure, and will probably continue to do so.

In my young days, we indulged in a bit of treasure hunting ourselves. We lived in Ben Hall country and the legend was that bold Ben had hidden some of his ill-gotten gains on or near our farm.

In one of our back paddocks was a steep rise known as "Ben Hall's Hill" and the story was that members of his gang had climbed one of the big trees at the top of the hill to watch the coaches take a back road route from the Turon diggings. They may have done that because some of the dead trunks that remained after we had gone in for timber-clearing were tall ones and would have given a view for miles around.

To back up the Ben Hall story were tales that after the coaches had been robbed, Ben and Co had hidden the gold somewhere on that hill, or in the country surrounding it, and had been killed or captured before they could return to claim it. There were plenty of hiding places among the big basalt rocks or in the hollow trees that abounded in the vicinity.

When I heard the story for the first time, I made a fairly close search of the locality, mainly in the hope of finding such things as discarded pistols, muzzle loading muskets and so on. But the results were negative.

As time went on, the old hands kept on with their stories of the hidden treasure until two or three of us combed upwards of the 150 acres on and around the hill carefully as a fosicker looked for specks of gold. We pried into everything, turned over rocks and logs, chopped out sections of trees and stumps and watched every rabbit warren within cooee of the area. For a time, we spent every spare minute we had on the search and during the periods we spelled, we discussed how we would share up the loot and what we would do with the proceeds. No-one bothered to tell us that the Government would have a claim on any treasure trove we discovered, knowledge that came to us years later when we had lost all interest in searching for Ben Hall's gold.

As you can guess, we did not find a thing, but the search was fun while it lasted. There was always the hope of finding something to urge us on, if we needed any urging, but the negative results eventually convinced us we were wasting our time. We turned to other, more profitable spare-time tasks, such as rabbiting, picking up dead wool and bones and sorting over potatoes.

Some time later the thought dawned on me that the old timers who had urged us on must have been at the same thing themselves when they were younger, and with the same results. They had a double issue of fun—when they searched themselves and when they incited us to do the same.

The Land, 5 July 1962

The terror that rides by night

There is, they say, a terror that rides by night, but to the young there can be several terrors and most of them are abroad after sundown.

My first definite acquaintance with this terror came when I was about 10 years old and it came through my favourite occupation of reading. I had got hold of a book, since out of print, dealing with some phases of early Australian history in which floggings, hangings, shootings, bushranging and violence generally were featured. Times might have been like it in the early days, but I doubt it because a small population like ours could never have kept going under such circumstances—violent death would have wiped them all out.

This book represented a new field to me when I borrowed it after church that Sunday morning. It was difficult to refrain from dipping into it before I went to tennis, but I kept away from the printed delights it offered me and tennis occupied the whole afternoon with pleasures of its own. However, as I was riding home, I looked with pleasurable anticipation at the book, knowing that I would have a couple of hours of uninterrupted reading after I had fed the calves and cared for my horse. By that time it was near sundown, so I lit the old kerosene lamp and settled down to an enjoyable read.

Other members of the family had gone visiting neighbours and I knew it would be some time before they got home. This was a good thing because it gave me the opportunity of reading without having to knock off for meals or to answer any of the numerous questions that might have been aimed at me. I curled up in a chair and started the book, which wasted no time getting into the violent aspects of the colony's early days. In fact, the descriptive narrative was good enough to make one's hackles rise even in broad daylight, so it can be imagined what effect it had at night time.

My first experience of the terror that rides by night came when one of the fowls squawked loudly. Cutting into the silence of the evening and across my preoccupation with the book, that squawk sounded like a hundred banshees rolled into one and it was some time before I settled down to the book again. I had barely got going when the beams started to creak in the house as the temperature changed. Anyone who has been in a house when that creaking starts will know just how terrible it can sound and if one has been reading, alone, about sudden death. It is definitely a terror that stalks abroad by night. Next interruption was not quite so bad when I became used to it, but it was bad enough when it started. One of our dogs began to bark and the bark was picked up by other dogs for miles around. Funny thing about dogs—sometimes they can begin barking and other dogs won't take any notice of them, but on other occasions the barking seems to be a message that must be relayed like smoke signals.

A few more noises, explained and unexplained, had brought me to a state of hair-raised terror, but I could not leave the book alone. Then came the clatter of the sulky horse's hooves, and never did a family get such a welcome home as mine did that evening. Their arrival brought an end to the terror that rides by night, to some extent, but I still had to face the nightmares that came when I tried to sleep.

The Land, 26 April 1962

Veteran shearer lacks confidence in champs

Old Harry was interested in the fact that champion shearers were to show their paces at the Sheep Show, but he doubted that any of them would be as good as some of the guns of the old days.

Harry had been a shearer for most of his long life until he retired, and he still shore a few sheep to keep his eye in, although they were pets of his children. One of his proudest boasts was that he had never learned to use a shearing machine—he had started with the hand shears and had continued with the "tongs" throughout his long career. This was not strange because Harry was a good shearer, greatly in demand amongst studmasters to shear their rams and the rates he received were well in excess of any award. In fact, Harry's announcement of his impending retirement caused more than a little consternation among the studmasters whose rams he had shorn for many years and with whom he had long been on friendly, first name terms. But he had trained one of his sons to do the job, which thus remained in the family, and the younger fellow soon consolidated his position, with the right words from Harry to employer and employee.

Harry always wore the old-time shearers' moccasins, the denim pants (definitely not jeans) and the heavy flannel singlet, known as a "Jacky Howe" and evolved originally to soak up perspiration and to prevent chills. He consumed large quantities of heavily sweetened black tea, and ate big meals of meat, vegetables and steamed puddings, with a few good whiskies in the evenings, at the station homesteads, where he was an honoured guest.

So, when Old Harry talked about shearing, he knew his subject, especially on shearing methods of a century ago, before the machines came into operation. He was a careful shearer and in his branch of the business there was no necessity to be fast, so I don't think he was ever the ringer of a shed, even before he got into the stud ram section.

"These chaps can shear with the machines, but how would they go with a pair of tongs?" he would ask. "It's all very well for them to have picked sheep, plain bodied wethers and ewes, for their demonstrations and for the New Zealanders to handle Romneys, but I'd like to see them on a shed of wrinkly, Vermont-type Merinos. That'd slow them up.

"And then they've got these experts to look after their machines and so on, but in my day we had to care for our own shears, make sure the leathers were right and sharpen the tongs to the right pitch, all a work of art. It would all be a mystery to these young blokes who call themselves champions. Everything is made too easy for them, as for all young people these days."

It was obvious Old Harry had never seen many machine shearers of ability in action, so one day we persuaded him to visit a big shed where the boys were racing for big tallies on difficult sheep. They worked like well-synchronised machinery and the wool billowed away from the hand pieces in great creamy-white waves, keeping the pickers-up and the classers busy. It was a revelation to anyone, even to our old friend Harry, and he had to admit that the "young blokes had something after all."

"But they wouldn't be too good on the big stud rams I've been handling," he said, dying hard, as usual.

The Land, 7 June 1962

Milking time had its moments of tenseness

There used to be a lot of debate about the merits of wet versus dry milking, but I don't think that any finality was ever reached.

In the days before milking machines, wet milking meant that the milker doused his or her fingers with a squirt of milk before starting on each cow, or dipped the fingers into the bucket after a few squirts had been taken from the cow. Dry milking, of course, meant that moisture was kept away from the teats. Wet milking was not really hygienic, although it may have been speedy, because drips from fingers contaminated the milk in the bucket and the dissemination of germs can well be imagined, especially as few milkers bothered to wash their hands after finishing one cow. Dry milking had its disadvantages, too. It caused chafing of the teats, accentuated trouble from small sores and scratches and sometimes resulted in teats being skinned.

Somehow or other cows seemed to thrive on either one or the other method, provided it was used from when they were broken-in to the bails and no change was made later. Whatever method was used, there was generally some trouble in hand-milking, no matter how well trained and conditioned the herd may have been. Some cows could be milked for months without a leg rope and then, one day, they would kick and milker and bucket would go sprawling—a waste of good milk.

Summer milking brought many such incidents and the leg ropes would be brought out. The cows resented the attention of flies in the bails. This was a double-edged business, as no milker could long tolerate being smacked across the face, generally over the bridge of the nose, with a tail as the cow attacked the flies that were pestering it. Under such conditions it was not long before the cowshed became almost a battlefield, with cranky cows and milkers rapidly losing their tempers. All parties were relieved when milking was finished.

Some cows had a vicious habit of using the foot nearest the milker to exact some revenge for disciplinary action against them. They would lift the foot, bring it down suddenly on the foot of the milker and give it a turn. Few pains are so excruciating as that, particularly on a frosty morning, and the milker could be expected to retaliate. Action against the cow, such as a kick in the ribs, or a resounding whack on the rump, had its repercussions, too, as she would "go off her milk," possibly for days. Generally being one of the best milkers in the herd, the loss of her milk would be felt in the total yield of the day. It would not always stop at one cow, as her mates would be upset as well, and they, too, would decline to "let down" their full quotas.

Sometimes it took days to get over such troubles, and the boss was always happy when cows and milker had settled down again.

The Land, 30 November 1961

Blackberry control is an everlasting task

Blackberries, one of the pests introduced to us by some enthusiast of a past generation, have always been troublesome in certain districts. They thrive on rough country and in cooler climates, and grow well elsewhere. It is in the rough country that the job of eradication is most difficult, even in these days of hormone sprays. It is there, too, that the blackberries provide harbour for rabbits. And despite myxomatosis and 1080 poison campaigns, there are still plenty of rabbits about which makes the job of cleaning up harbour all the more important.

Many methods of getting rid of blackberries have been tried with varying degrees of success. A favourite was to slash the canes down with a bill hook, allow them to dry, then fire them. This treatment was followed with ploughing and sowing with good pasture grass. Allowed to remain at that stage of treatment, the blackberries would thrive and it would not be long before new and healthy shoots made their way through the soil. Then the most important phase of the campaign began. The blackberry plot would be fenced and a mob of hungry pigs or goats enclosed in the area. They would soon eat down the grass and would then start on the blackberries, which seemed to be tasty morsels, especially to the pigs. Generally the pigs did a better job than the goats or any other livestock because they would root up the soil in search of blackberry roots and other succulent growth. If the pigs got no other food for some days, hunger would force them to do a particularly good job on the blackberries and their activity would improve the soil. By then sowing a fast growing crop, there was a chance of smothering out a big proportion of the surviving blackberries, but it was impossible to get rid of them entirely.

Further treatment, including ripping up, heavy grazing and smothering, would be necessary for a few years. If the work was relaxed for any length of time, the blackberries would grow more vigorously than ever. This treatment, of course, was applied to the "easy" country. It was anybody's guess how the pest was to be tackled on rough country. Slashing, burning, hoeing, all had their part, but the work had to consistent. Sometimes it succeeded, sometimes it did not and blackberries were a constant worry.

Modern methods, apparently, have brought some relief, but blackberry control is still not a lazy man's job. From what I have seen the sprayed clumps of canes will grow again in a year or two, regenerated from the roots. Constant work will keep the pest in check but there can be no relaxation of effort.

Even when one is always vigilant, there is a chance that the blackberries will win in the end. I found evidence of this when inspecting a splendid lawn in a long settled area. That lawn had been fertilised, watered and cut regularly until the holiday period when no mower had known it for a little more than three weeks. Growing up through the grass were healthy blackberry shoots ready to take over if the neglect of the lawn continued. And it must have been at least 15 years since the lawn was established. So, the war against the blackberry must be considered a lifetime assignment.

The Land, 22 February 1962

At first show

In between the annual Parramatta Show and the Sydney Royal Show is an appropriate time to recall the earliest days of showing in Australia. In fact, the first show. This was held at Parramatta, then the near capital of the penal colony of New South Wales, in the early 1820s.

My main interest in that show comes from the stories about the first show my grandfather heard from his father. My great-grandfather was only the broth of a boy, but he was observant and he had a retentive memory, which, with his talent for embroidering his tales of the early days, made them particularly interesting.

Because there was no room at the inns, it was necessary for some families to camp out. They erected their big tents and had an enjoyable time of it, grilling their meat on big fires of coals and making buckets of tea. They spent most the night drinking and yarning. All efforts to get the young 'uns to bed failed because of the strange surroundings and the excitement of the show. This was to be unlike even the fairs that the non-natives could remember in the old country. But to the young cornstalks who had seen nothing like it, the show promised untold delights.

Dawned the great day and from an early hour people began to converge on Parramatta by foot, horseback, in coaches and the common dray. From Sydney came the quality—many of the men in the uniform of the regiment then providing the garrison, others in Naval uniform, and the ladies, under gay parasols, in the most fashionable crinolines. A somewhat rougher type came from the settlements around Parramatta and on the Hawkesbury, but they had the glint of independence in their eyes. Even the Sydney-siders knew they were the people who grew the food for the colony, keeping it from the semi-starvation it had known so often. They had a right to be independent, my great-grandfather considered.

Then, of course, there were the convicts who attended the show with their masters and the blackfellows who wandered around on the outskirts, amazed at this latest example of silliness on the part of the strange white people who had come to their country.

Being a good year, the show attracted a grand display of exhibits, the best collection ever put together in the colony to that date. If it did nothing else, the show convinced many people that New South Wales had a future, and a future that would be much brighter as the new lands over the Blue Mountains were more fully developed

But all these thoughts came to my great-grandfather in later life. Somehow or other he became involved in a senseless argument with a young fellow from Sydney Town whose father apparently was an officer and who had been taught something about boxing. My great-grandfather knew nothing about boxing, but he did know quite a lot about fighting of the kind he had picked up from convicts working along the river and from some of the tough characters who could fight as well as they could farm. He emerged from the clash that occurred as the actual victor, but he was in poor shape. So poor, in fact, that his mother could not refrain from weeping and his father declared that he did not have the heart to give him the belting he deserved when he was in such a state.

The Land, 12 April 1962

When the bonny pipers roused Scottish blood

Funny how the bagpipes affect different people, in different ways—they arouse a spirit of pride in some, martial longings in others, and the desire to commit murder in some more.

Being of Scottish descent on my mother's side, I have always had a liking for the pipes and once joined a pipe band, although I never reached the stage where I could be classed as a competent piper. One thing against it was that I was in my teens, when there are a lot of things to do rather than spend hours practising with the chanter, an essential operation if you ever hope to conquer the pipes.

This chanter is a whistle-like contraption, fitted with the same reeds that make music in the pipes, and holes on which you use your fingers to get the tune. Once you master the chanter, however, you can say that you are on the way to becoming a piper. All you have to learn then is to carry the pipes properly, pump the bag with your elbow, keep the bag full with a good supply of air from your mouth, get the drones and the chanter itself in tune, and then go ahead. As you will see, playing the bagpipes entails a lot of thought and you have to synchronise your operations, using your fingers to let the air out as needed to make the different kinds of noises.

It is no use anyone saying that one tune played on the pipes is much like the other—a study of what is produced by a piper will show there is a vast difference between, say, *The Barren Rocks of Aden* and that beautiful lament, *The Flowers of the Forest*.

As youngsters, we were given a strong dose of bagpipes at least once a year when we all attended an annual Highland Gathering. As this was held 30 odd miles from home, it meant getting up early and doing the odd jobs around the farm so we could get behind the speedy trotting mare we drove in the sulky and then off to the Gathering.

That was always a feast of entertainment, yarns and fights for some, a fact that made a special appeal to the Irish side of my ancestry. There would be footraces, tug-o-wars, tossing the caber, putting the shot and other forms of vigorous exercise for strong men, many of whom competed in the kilt. However, one of the main attractions was the Highland dancing, which went on through the day, competitors ranging from tiny tots to bearded ancients, all in full Highland dress, which must have been pretty hot as the gathering was always held in summer. For each dance, a piper would play the tune and the kilts and sporrans would swing in tune as the nimble-footed dancers went through their paces. Some of the men would be well fuelled up with whisky by lunchtime and this seemed to improve their dancing.

At odd points around the ground, especially near the bars, pipers would be blasting away on solo efforts with great vigour, the main intent to drown out the sound of the nearest rival set of pipes. At times, the pipe band would be called together to stage a march, sometimes with the kilted dancers and others in similar attire marching behind as one could imagine their ancestors doing in the wild highlands of Scotland.

For anyone in whom the Scottish blood was strong, the Highland Gathering was a day to enjoy and remember.

When everything was over at the gathering, the assembled Scots who were sufficiently capable would form up behind the band and march down town as publicity for the concert to follow after all had been fed and somewhat rested.

On one occasion the band, well seasoned with the national drink, was playing down the street as if it was taking a Highland regiment into action, when a large, fightable looking character made a caustic remark to a diminutive drummer. He did not know that the drummer was an able fighting man and that the four other drummers were just as good with their fists—and the five were brothers. As one man they dropped their drums in the roadway, dashed across to the man who had issued the insult and had him knocked out with great speed. Picking up their drums, they trotted to join the band ahead of them, dressed by the right and were into the tune as if they had not been interrupted. And the pipers played on, apparently unaware that they had been without the aid of drums.

The Land, 31 May 1962

First day at school

One's first day at school is something to remember, but it is always a little more memorable if the school has just been re-opened after being closed for some years. On such occasions there are always some youngsters of eight and nine who have never had any experience of school and the disciplines it imposes. They are old enough to query some of the stories put out by the more senior pupils who are not taking too kindly to the re-imposition of normal discipline. Round the eight and nine mark, little notice is taken of the old saw that "a still tongue makes a wise head," and remarks are made that will get one into trouble sooner or later.

Our schooling began under such circumstances. I was just on eight and apparently precocious for my years, which did not endear me to the older lads, nor to those who were younger and who had brothers older than themselves. So it was natural that I was due for taking down a peg or two.

It all started before school opened when suspicious groups of youngsters clustered around eyeing each other off. We got tired of that and some of us moved to the top of the school ground, beyond the tennis court where we were followed by two or three of the big chaps.

My mother had decked me out in new clothes, topped by one of the old straw boaters, a form of headgear that placed temptation in the way of my bigger colleagues right from the start. Their powers of resistance against such temptations were nil, and it was not long before my boater was snatched off my head and sent bowling down the hard surface of the tennis court. As it spun around, another of the lads raced down and recovered it, to send it spinning back to the original joker. My protests were of no avail and they kept themselves happy spinning the hat along the ground until the brim had gone. By this time I was close to tears and had not recovered when the bell pealed to call us into some sort of order.

A brief address from the teacher and we were herded into the schoolroom where we were mustered into tentative classes prior to starting work. Long ages seemed to pass before the morning break came. For one accustomed to an open-air life, confinement in a schoolroom was as bad as going to gaol, but I did not realise that I would have problems when we eventually did get out of school.

No sooner had we reached the playground than the bigger chaps decided that another lad of about my own age and size should fight me. We had nothing against each other, therefore the preliminaries took longer than had been anticipated and we were back in school again before we had our coats off. At lunch time it was a different story. The fight promoters got on the job again. This time they worked us both up to such an extent that we began throwing punches and one connected with my nose. That made me see red and I launched myself upon my reluctant opponent. I don't think any of my punches connected, but I pulled him to the ground and was going to work on him when we were brought to a sudden halt by an adult voice. Looking up we saw the teacher who reprimanded us for fighting and threatened to keep us in after school.

And where were the chaps who had got us into this mess? Not one of them was in sight—they had stolen away to other parts of the school ground, leaving us to take the rap.

That fight was never finished because we became good mates and would merge our forces to defend each other from outside attack.

A long afternoon eventually came to an end and we wended our way home. Then I remembered—I had forgotten to eat my lunch. I sat down with my back to a stump and got through it. It was the best event of the day.

The Land, 5 October 1961

Bird Day—a big occasion

Bird Day was a big event when I went to school, not so much for the acquisition of knowledge about our avian friends as for the associated activities.

Before we all became members of the Gould League of Bird Lovers, the big thing was to make collections of birds' eggs. The school itself had an excellent display, to which all of us contributed eggs we did not want ourselves. That sort of thing came to a sudden halt immediately we joined the League, whose main objective was to preserve bird life. For a small fee we became proud owners of League badges which we wore until we lost them. If I remember correctly, we had to make a promise to do all we could to ensure that birds were protected and encouraged to increase in numbers. That kind of thing took some of the excitement out of life for the conscientious members of the League.

Prior to League membership, however, Bird Day had its thrills.

We attended school but after roll call the teacher would lead all pupils on an expedition into the bush. We'd be laden down with lunches and a couple of big billycans for tea-making, and the cigar boxes in which the eggs we collected would be stowed. In those days our principal aim was supposed to be the observation of bird habits and the study of their nests, but we gave the nests first priority.

Straggling through the undergrowth, boys and girls alike would be searching for nests. Shouts would indicate that someone had made a discovery. If the nest was in a tree the height above ground would govern the choice of the lad who was to climb it and take one egg for the school collection. Girls were allowed to take the eggs from any ground nest found. They got few opportunities as only an occasional lark's nest would be found, except for one red-letter day when we flushed some quail.

Our best climber was, like John O'Brien's hero from Tangmalangaloo, "an overgrown, two-storey lad," and he was appointed to climb the tallest trees. On one occasion he did an excellent job of shinning up the bole of a gum tree with no branches to help him for about 15 feet. Warned by the teacher to be careful, he went out on a limb to a magpie's nest, from which he extracted an egg. Holding it in his fingers, he asked: "How'll I bring it down?" He had neglected to wear his hat.

A hat would have simplified this task because our favourite method was to place the egg under the hat, our usually generous growth of hair providing a safety cushion for the fragile shell. We yelled advice at him to put the egg in his shorts pocket or in the front of his shirt, but he rejected such methods as being conducive to breakage because of the possible jars and jolts as he slipped down the tree.

Eventually he decided to carry the egg in his mouth, a style new to him and to most of us. Dubiously, he put the egg in his mouth and began his descent. All appeared to be going well until he neared the ground. His grip on the bole of the tree slipped and he finished the journey at accelerated speed. Immediately on landing he began spitting out eggshell and yolk. "The ——— thing was rotten," he gasped as he turned away to be violently ill.

The Land, 20 April 1961

The half-time school

Are there any half-time schools these days? If not, the shortage of teachers may encourage the Education Department to resume the practice it followed a generation or so ago in many country districts.

My early schooling was on a half-time basis. The system worked this way: a teacher had charge of two schools, several miles apart, neither of which could gather enough pupils to warrant five days a week teaching in the one building. One school would be open for business Monday, Wednesday and Friday this week, Tuesday and Thursday next week. And the other school was used to impart knowledge on the alternate days.

Teachers didn't like the system, but had to put up with it. Parents shared their dislike for various reasons, one being the break in the teaching with the accompaniment of homework that senior members of the family were supposed to supervise.

Our first teacher was an attractive young lass, daughter of neighbours and fresh out of college. Some months after school began, a frequent caller was the widower parent of three of our schoolmates. "Miss", as the teacher was known, would leave the schoolroom (our place of learning boasted only one room) when he called, and return to us some time later with hair out of place and a flushed face.

This puzzled her innocent flock, but was inevitably reported at home. We could not fathom the knowing smile with which Mum greeted the news, but we did not care much. However, it was not long before we got another teacher—"Miss" married the widower and retired from the service to raise a brood of scholars of her own.

The Land, 13 April 1961

When fowls were run primitive style

In the days when fowls were allowed to run an extremely wide range, and the eggs represented additional income for the housewife, quality was not always the main consideration.

Frequently, the fowls were allowed to return to their natural state and to roost in trees, sometimes given auxiliary perches in the form of a few slats nailed to a couple of poles, something like a wide ladder with rungs well apart. This natural state roosting was all right because the fowls were able to get out of the reach of foxes, but there were other predatory enemies. These included native cats and domestic cats gone wild and they could wreak havoc at night.

Modern day poultry fanciers would look askance at those "houses," which were knocked together with any available materials, generally against a fence and with a bark or galvanised roof. A few poles were used for perches and everyone seemed happy.

Little attention was given to breed. Some of those old-time farm flocks represented a multitude of crosses that would have perplexed a specialist in genetics. You'd find big and little hens of all colours, but they were usually run with a cock bird selected mainly for the beauty of his plumage. "Selection" was no specialised task either—he'd be picked from a clutch of chickens for his colour and size and the other cock birds would go into the pot.

Naturally, with such a mixture, eggs would cover a big range of sizes, from bantam grade to double-yolkers. But they were all useful in the kitchen and any surplus could be sold to the store in town, or used as a contra against the grocery account.

Finding and collecting the eggs was more a game of chance than one of skill. The hens developed some cunning and would make their nests in all sorts of odd places, sometimes half a mile or more from the house. Finding those nests became an exercise in bushcraft and scouting for the younger members of the family who would trail the hens, sometimes for days, until they found the eggs. Usually each nest would have a dozen or more eggs, stained with yolk from a few that had been broken, and, generally, dirty. However, the whole lot would be collected.

Then began the task of sorting the eatable from the non-eatable. This was usually done by breaking each egg into a cup, and if it did not smell it went into a basin for use in cookery. In summer, each nest of eggs would contain several that definitely were not edible—their aroma when cracked signified that.

Once a nest was discovered, more or less regular collections from then on ensured a certain degree of freshness. But all of us were not so careful with the eggs we supplied to the grocer and it is to be feared that many town buyers were dissatisfied with the quality of the eggs they bought. I have no idea how the grocer got over that hurdle.

In running fowls under such conditions, hens hatched out their own eggs and raised the chickens, as incubators and brooders were unknown. Frequently a hen would arrive back at the house, generally at feeding time, with a brood of chickens around her. She had hatched them out away from human eyes and interference. But management of the farm flock has improved since those days. Some of the descendants of the early settlers now breed prize-winning birds, not, it must be understood, descendants of the early day layers.

The Land, 15 June 1961

When summer water was from "The Spring"

When the tank water supply failed during a hot, dry summer, as it always did when I was a boy on the farm, it became necessary to cart water for household use. In some ways the job was a pleasant one, but in others it was onerous, especially if rain was delayed for any length of time.

Our water-carting equipment was not extensive and rarely could we cart more than was used in a day. Some of the youngsters endeavoured to eke out the supply by refraining from taking baths but that neglect was soon discovered and the offenders were given an impulse to remember with a clip in the ear. We carried water in a large barrel on a dray. The barrel was surrounded by cream cans of varying sizes, kerosene tins and buckets.

Our emergency supply came from "The Spring," a never failing supply on the side of a hill about half a mile from the house. We were lucky to have "The Spring" because some of our neighbours were forced to travel miles each way when carting water.

Our spring bubbled from the side of the hill into a hole that had been carefully excavated and from which buckets full of water were transferred to the barrel and other containers. No interference with "The Spring" or its surroundings was tolerated because it was feared that any changes or excavations would stop it from flowing, as had occurred with other springs in the district. Surplus water was allowed to trickle from the hole in front of the spring into channels that provided drinking water for stock and irrigated a portion of the flat that rolled away from the hill.

Before water carting was due, the old barrel would be trundled out of the shed that had housed it for most of the year. To tighten the staves, it was wrapped in wet bags until it had taken up sufficiently to hold water. This water ran out in gradually decreasing quantities over a few days until the barrel was declared watertight. It was then roped to the dray where it was draped with wet bags. After loading all vacant spaces in the dray with other containers, we would be off with our old dark grey horse between the shafts.

At the start of the job we usually carted the water early in the morning, but the time was variable and was affected by other, more urgent, work that had to be done. One of us would dip the spring water and hand it to another member of the team on the dray, who would empty it into the barrel. It could be hot work, despite the cool surroundings of "The Spring" but it was a task that had to be done. You could always stop for a drink of the cool water as it came from the spring, water that had a distinctive mellow flavour of its own. It lacked the sharp tang of tank water and the distinctive flavour of well or river water, but it was good. The women used to tell us that it was "soft" for washing clothes, probably because there was no limestone in the district. Loaded up, and with a handful or two of watercress for salads or sandwiches, we would travel back to the house, water dripping into the dust of the track. Our dray would be parked under a tree in the yard near the back door where the shade helped to prevent undue evaporation.

Carting water, after a week or two, became arduous and no one complained when the rains came and the need to continue the job ended. But we were lucky to have "The Spring", which often proved a boon to some of our nearest neighbours, too.

The Land, 25 January 1962

Billy tea must be black and strong

A cup of strong black tea, liberally sugared, is hard to beat at any time, but is at its best on a hot summer day. Generally, the tea tastes better if it is made in a well-blackened billy can, boiled over a fire of gum sticks and with the lid off. This enables some of flavour of the burning sticks to be imparted and gives it what a wine expert would describe as the "bouquet of the bush."

To make tea of that type, you bring the water to the boil, throw in what you judge to be sufficient tea leaves, and put the billy back on the fire until it boils again. Then you serve the tea, preferably in tin pannikins, but it must be hot, especially on a hot day. Drinkers add their sugar to taste and it is generally a liberal addition. Some people prefer black tea without sugar, which they reckon cuts all the dust and other matter from the throat, as well as proving more stimulating when the need is greatest. They are tolerated by the billy tea addict, who normally has no time for the chap who spoils the brew by adding milk. To the hardened drinker, the addition of milk spoils both tea and milk. He prefers to take both on their own.

There are some tricks in making billy tea. For instance, you can prevent the water from being smoked if you lay a sturdy green gum twig across the open top of the billy. Tapping the side of the billy with a stick is advised to settle the leaves when the tea is made. Sometimes the tap needs to be a strong one; at others a few gentle taps will do the job. Some tea makers settle the leaves by grabbing the handle of the billy and swinging it around. This is all right when you learn how to handle the billy, but if it is swung too much the tea cools down.

Much depends on the water used to make the tea. All water available in the bush has a flavour of its own, ranging from the iron tank in which birds have drowned to the lazily flowing river in which you might find some dead sheep after you have had your tea. You'll have to make tea out of muddy water sometimes and use sparkling clear water from the snow country at others. It is probably because of the varying water flavours that the bush tea maker believes in strong tea which disguises the flavour of the water.

To really enjoy a mug of billy tea, knock off at "smoko" on a hot day after wielding an axe, pitchfork or a pick, and down a couple of mouthfuls. After that you would not take nectar from the gods. Under such conditions many bushmen agree that tea is better than beer, although they will argue that each drink is good in its own place.

A pannikin of good, hot, strong black tea, well sweetened, opens the pores and makes the perspiration flow, which seems to help when you are working under conditions that might well be applied to the poet's "Hay, Hell and Booligal". There's nothing like good billy tea—but it must be made in the right way.

The Land, 15 February 1962

Many a good cook had to start with a camp oven

Driving along a little used road in our neighbourhood the other day, I noticed a camp oven sitting on a post. There was no apparent reason why the oven should have been on the post, and by the look of it, that post had been its resting place for many a long day.

It was obvious that it had not been used as a mail box and it was too small for a bread box. Evidently, someone had stood it on the post at the front of the house and it had stayed there because no one wanted to use it again. One surprising thing about that camp oven was that it was capped by a lid that seemed to be in good working order.

When I was younger, camp oven lids were always a problem. They were always getting broken or lost and various stopgaps were used to replace them. But you could never really cook well in a camp oven unless its own lid was used because it fitted properly and prevented any loss of heat. Our camp oven had legs that enabled it to be set amidst the coals, and for other coals to be raked up around it, either in a fireplace or in a sheltered spot in the open.

Possibly, the camp oven contributed more to pioneer living than anything else. It was the only approach to a stove or oven that many of the pioneer homes possessed and it was an ideal vessel in which to make damper, and even bread. The camp oven was unexcelled for baking meat and vegetables—even by some of the early fuel stoves. With the coals raked all over the oven, and potatoes in their jackets placed on the lid and then covered with more glowing coals, the cast iron of the oven would absorb the heat. It would cook a meal the equal of anything you could get, even today, at a leading hotel—always providing the cook knew what he or she was doing.

Our camp ovens were used at home mainly as curiosities, or if someone wanted to show off their bush cooking ability. It was an accepted thing in those days that the girls be taught camp oven cooking as well as the use of the ordinary stove. We used our ovens when we went on long droving trips, camping or excursions that took us away from home for any length of time. My first lessons in cooking dampers and shearers' brownies were with the camp oven. I used the same method when I was taught how to make a sea pie, using mutton or beef, or a burrow pie using rabbit.

I suppose if you showed a camp oven to most Australians these days they would not know what it was, nor what it was used for. Yet it is not so long ago that the ovens were in fairly general use in the pioneer districts. Many things have been cooked in camp ovens, and cooked well, and some of the early families stuck to them long after their neighbours had bought stoves. Such an oven-loving family would have a battery of perhaps half a dozen of the squat little iron receptacles. They were thus able to cook a meal of several courses for a big influx of visitors, or to bake a batch of bread that would feed a big family and the employees for upwards of a week.

Some of the housewives I knew when I was a boy were unable to bake bread in anything but a camp oven—they had become so accustomed to that style of cooking. On the other hand, there were housewives who smashed their camp ovens with the back of an axe immediately a stove was installed. It all depended on the point of view.

The Land, 10 May 1962

Modern Mrs would balk at pioneer washing day

Equipped with a modern washing machine and a rotary clothesline, housewives of today, especially youngsters, would have no knowledge of the conditions under which their immediate female ancestors did their laundry. Of course, there are women throughout Australia unable, for various reasons, to have modern methods of washing, but their numbers are declining every year.

Even less than a century ago the common laundry equipment comprised a couple of galvanised iron tubs, a copper, a washing board and a supply of home-made yellow soap. That equipment was standard. Recently I was reading some historical references to the now wealthy Western District of Victoria where the washing was done in the squatter's homes in the same primitive manner as that by the station workers' wives for their families. This had been going on all over the various colonies since they were first settled and many of us can recall the laundry set-up of our own homes when we were young.

The washboard, or scrubbing board, was a piece of equipment that always fascinated me. These boards were made of rounded slats of wood fixed in a frame. The whole thing stood in the tub where the clothes were sloshed up and down in the water, soaped and rubbed against the slats at right angles. Later on the wooden slats were replaced by narrow corrugated galvanised iron on some makes of boards. These were longer lasting units that did not splinter with use.

Not much can be said about galvanised iron tubs, except that they came in various sizes and were used for lots of other purposes between weekly washes. They served as baths, as a receptacle for treating wheat before sowing, something into which to cut potatoes, and even in which to make home-brew, although they had some effect on the taste of that. If big enough, a tub was ideal for scalding a pig to enable the bristles to be scraped off and most farmers made sure the laundry was equipped with tubs of a suitable size for this job.

Coppers were generally built in with mud, stone or gravel in the early days, and later with brick, to conserve the heat and ensure early boiling because they, too, were used for more than the weekly wash. If a hot bath was wanted, the copper was stoked up; it was boiled for tea for the harvest hands or the shearing team or when there was any invasion of additional people for large quantities of tea. It was boiled to provide water during pig killing, for scrubbing purposes and for the hot water necessary to keep the dairy clean, although the authorities were not so fussy about cleanliness in those days.

Even now you'll find a copper and a tub or two in use on many farms that are equipped with washing machines. But housewives don't use the old-time equipment in the conventional manner—it comes under the heading of the male side of farm plant. Except, of course, at Christmas when the copper is handy to boil a big fruit pudding, although the mixing is not generally done nowadays in a tub.

Then, of course, there was the soap, sometimes bought, but more often home-made by the earliest settlers. I can recall different types of soap that had been made at our place and at others, all varying according to the recipes and skills of the makers. This soap ranged in colour from a ghastly grey looking mass to a golden yellow, the equal of any of today's top brands turned out by the big companies. Some forms of soap flakes were on the market in my youth, but they were considered expensive except for the finest materials and it was long before the modern powders and flakes caught on.

But merchandising methods, advertising and the advent of the washing machine have revolutionised that side of home life to such an extent that the young wife of today would balk at a slab of her grandmother's home-made soap, to say nothing of the other laundry equipment.

The Land, 12 July 1962

A clothesline could often be a menace

Rotary clotheslines are making an impact throughout the country following their almost complete takeover in the city, town and suburban areas. But the old style line is still to be seen around many homes. And the country clothesline varies as much as the individual who erects and uses it.

One of the originals that still finds favour is the length of rope stretched between two trees with a light, forked sapling as a prop in the centre. At the other extreme, you'll find a double line with the wires strung neatly on an arm of sawn and planed timber at each end, the arms being bolted to a sturdier sawn and planed upright. The aim of this type of line is to permit the arms to swing up and down. When the line is full, the arms are tethered to the uprights, usually with something like a dog chain, to keep the wash on an even keel while it dries.

Even with this, the immediate predecessor of the rotary hoist, trouble can come on washing days. The wires can break after some years of use. The tethering chains break and the wash falls to the ground, much to the ire of the laundress who, when the men are about, speaks her mind with freedom on lack of maintenance.

Accidents with the more primitive type of line are more common because the line itself, whether rope or wire, is likely to break or come undone at the ends. Another danger is that the prop will break, especially on a windy day, casting the wet clothes into the dust, mud or grass, depending on the season.

Sometimes the woman of the house looked after the maintenance of her own clothesline, but I was never lucky enough to be a member of such a household and the job was always mine. And many the harsh word have I heard when a washing day accident occurred; many a time have I jumped to repair the damage as the clothes were put through the tubs again.

Clotheslines could cause other trouble, too. Someone would leave the prop down at night and someone else would walk into the line in the dark and be thrown heavily to the ground with resultant curses and other bad language. But the most painful of such accidents was when you rode a horse into the clothesline, especially at night. It would come as a shock to hit the wire anywhere from the forehead to the waist and have it drag you from the saddle, back over the rump of the horse. It would not be until you hit the ground with a bump that you realised what had happened to you.

Until the more or less recent introduction of plastics, which followed the use of springs with smaller wooden pegs, the design of the clothes peg does not seem to have changed for generations. Many millions of the old cleft wooden pegs with the rounded top must have been made over the years and made fortunes for some people. This I know, because an acquaintance of mine married, somewhat late in life, a clothes peg heiress and never worked again. They spend their time travelling Australia and the world and obviously still have a fat bank balance left which makes one realise that even the simplest household needs should be treated with respect.

The Land, 19 July 1962

Magpie's nest was wire work of art

One of the prized possessions of our school was a magpie's nest constructed entirely of tie wire. It was built during the period when most of the surrounding farms were being wire-netted and one of the pupils brought it along to school. The nest was displayed in the small museum at school and was taken out for inspection by visitors.

The birds showed ingenuity when they built the nest: short lengths of wire had been interwoven in a lasting fashion. The lining was more conventional, being of hair and feathers. If that nest had been allowed to remain in the tree where one lot of young birds had been hatched and reared, it would probably have lasted for several seasons as it was the most indestructible magpie's nest I have ever seen.

★   ★   ★   ★

Browsing through an overseas farm journal, I came across an interesting item which indicated that people in other countries had not been taught the lessons that some of us Australians had learned.

This item stated: "No one knows why, but animals of the ruminant order (cows, sheep, goats, deer, camels, giraffes) when they are down, get up on their knees, or hind feet first; other four-legged animals get up on their front legs first."

Long ago I was told that the ruminants, armed with horns, rose hind feet first for reasons of defence. As they levelled themselves out, as it were, they were in an attacking position, with the full weight of their bodies behind the thrust forward at an enemy. Similarly, other four-legged animals rise on their front legs first to give them purchase to get the heels of the hind legs into action.

It is all supposed to be a relic of past centuries when present-day domestic animals owed their very existence to horn, heels, tooth and claw.

★ ★ ★ ★

Burrcutters may be a vanishing race, but they, or their modern equivalents, could be employed to advantage on some of the wheat lands of the Riverina, South and South-west, at least.

Travelling through those parts just after the last harvest, I was amazed to see thousands of acres of various species of burrs. It must be difficult for some of those growers to market clean wheat.

Obviously, the burrs have been given an open go and, unless they are checked, will take possession of the paddocks. In a way, the burrs could prove as serious a pest as skeleton weed. Their destruction offers scope for chemical and mechanical methods—and the old-time burrcutter may yet ride again.

★ ★ ★ ★

In the early days of settlement, houses were built of wattle and daub—roughly hewn slats nailed as a "box" and filled with mud. Some of these houses are still standing, although they are more than 100 years old.

Later came the pise method of construction, walls of nothing else but earth, and the houses so built have given good service for many generations. The earthen walls, smoothly plastered inside and whitewashed outside, provide insulation against extremes of climate. It is a cheap method of construction and, in these days of high building costs, it often puzzles me why more pise houses are not being built.

Some years ago, efforts were made to arouse interest in pise construction on the Murrumbidgee Irrigation Area, but how successful it was I do not know. In other countries, including Spain and the South Americas, pise buildings have long given good service.

The Land, 29 June 1961

Willows for beauty sometimes trouble

When winter comes, sometimes much of the beauty goes from the willows that fringe many of our streams. But their bare branches give an old-world atmosphere to the countryside and a close watch tells us when spring is coming. It is then that the fresh buds appear, often before we realise that the winter is almost over.

Some of those willows grow to a tremendous size and their branches cover many square yards of space. Swagmen have been known to make their homes under the overhanging branches and to live there until winter stripped the leaves from the branches. Our swaggie friends would be snug in these tree homes, into which they would introduce many of the limited comforts of their ilk. However, a sudden flood or a bushfire would bring a rapid exit.

Willow wood might be ideal for making cricket bats, but it had few uses on the farm and as firewood it burned too quickly.

Willows add beauty to the stream along which they grow, but they hide things, too. For instance, you might come to a river panting for a drink and with your horses in the same condition. As you gulped the cool and refreshing water you blessed the stream and the trees that helped to keep the temperature down. After a spell, you might decide to venture upstream to see what the country was like, whether any gravel was available, or just to admire the scenery. And you would be shocked to find the decomposing carcasses of a couple of dead sheep in the water, washed by the stream in its downward progress. That has happened to many of us and, despite modern ideas of hygiene, we are still hale and hearty.

Some people like to have a willow tree or two growing around the place even if they are miles from a creek or dam. My view is that the trees should be planted around dams, on flats, in swamps, in fact anywhere well away from drains and wells.

Nothing is calculated to make a mess of drains as efficiently as willows do. Their roots find their way into any chinks in search of the water and before long they grow into big balls and extend their tentacles in both directions. Suddenly the drain ceases to work and waste flows all over the place. A major operation is then essential, but it is no good just cleaning out the roots and leaving the tree stranding. If you do that the drains will be blocked again before long as the root pruning seems to do the willow system a lot of good. Therefore it pays to grow the willows well away from drains and pipes despite their beauty and the splendid shade they provide in summer.

Still, taken by and large, the willow has proved an asset to Australia. Their distinctive green adds to the beauty of the landscape, especially when you are on a rise and view the trees along the course of a creek or river.

The Land, 14 December 1961

When kerosene lamps lit our darkness

Electricity is common in most districts of several States these days, coming either from a public authority or from a home generating plant. This is one mark of progress that has been made in Australia in recent years and it has happened so gradually that it has been taken more or less for granted. But many of us can remember when few of the bigger towns were lighted by electricity and when home-generated supplies were not available, even to the wealthiest of homesteads.

In fact it is not so long since the days of the kerosene lamp and, in the comparatively brief life of Australia, since the times of the of the home-made candles and the slush lamps of the pioneers. You can see the primitive equipment used for making candles in museums, but you cannot experience the "high" odour of the candles as they burnt. A slush lamp was even worse—as youngsters we had heard a lot about them and, following the directions of our grandfather, made a couple. We filled a shallow tin with fat, twisted shredded cloth around a splinter of softwood to act as a wick, and lit it. It was necessary to first melt the fat that the wick would draw up and keep burning with an attendant stink. That meant our slush lamps were out of the house in double quick time because Mum reckoned she'd never get rid of the stink. What was good enough for the pioneers was not good enough for our mother.

Kerosene, or paraffin oil, lamps followed the candles. Every home owned a few of varying sizes. They provided all the illumination needed for the home, even for hours of reading and study by youngsters who were ambitious to get on in the world. When reading under modern fluorescent lights it is a marvel to think back to the kerosene lamp days and to realise that we did not all go blind as a result.

Living in an area subject to blackouts whenever we have a storm, we have acquired a kerosene lamp that is regarded by the younger generation as a curious relic of the days when Dad and Mum were kids. They like it on the table when meals are eaten, apparently for its romantic golden glow, but they refuse to read by it as they say they cannot see. Give them the electricity to which they have been accustomed from birth, every time.

For safety sake, we always kept a spare globe on hand as such fragile equipment broke easily. Wicks had to be trimmed or they would smoke and heat the globes with resulting explosions. Cleaning the globes was one of the household jobs.

We bought kerosene in tins, which were useful for making buckets and various other utensils. We generally bought two at a time which meant that we got a box as well, and that had many uses.

Many people were scared of the first power lamps when they first came on the market. You pumped these lamps up to get the pressure. They shed a bright white light that was a vast improvement over the old kerosene illumination, but the pressure was the scary point. They caught on, however, and pressure units were installed on many places to serve the whole of the home from one central tank and pump.

Not long afterwards the home electricity-generating unit came on the market and enjoyed brisk sales. It was followed by the extension of supplies by public utilities. Today electricity lights more farm and station homes than anything else.

The Land, 14 June 1962

Winter reading circle around the log fire

Despite half-time schools, or no school at all lots of the time, the members of my family were brought up to have an appreciation of books and their contents, even if some of the material would not be approved by present-day experts on education.

During the long winter evenings we would stoke up the big open fire in the kitchen with a couple of logs and some small wood to keep it burning and gather around in a circle. Dad would sit nearest the kerosene lamp and read to the rest of us.

As I remember, we started off with *David Copperfield* which gave us an advantage over our schoolmates when that book came up for study at school: we knew all about it. Other volumes dealt with by the reading circle in its early days were *Robinson Crusoe* and *The Swiss Family Robinson*. These, too, were the subject of later study at school and kept us in front. G.A. Henty, Fenimore Cooper and Ballantyne were other popular authors, and we went through Louisa M. Alcott's *Little Women* and *Good Wives* in due course.

However, we gave more than a little attention to Australian writings, including a series on Australian bushrangers, such as *Ben Hall, Bushranger* and *The Kelly Gang*. Of more than passing interest were books written by J.H.M. Abbott and covering some of Australia's earlier history. Many of these works had a tremendous impact for us because members of our family had lived or worked in the districts in which they were set. Our grandfathers had been associated, in one way or another, with some of the characters including the bushrangers in a hostile sort of way. We'd been able to recall the stories they handed down, some of them better than the authors had included in the books.

But the really popular works were those of Steele Rudd: *On Our Selection, Our New Selection* and so on. We went through the lot of them in community and individual reading, until they showed signs of wear and had to be stuck together for future readings. The popularity of the Rudd books was possibly due to the fact that the conditions of our own existence were not so far removed from the people they featured. They were close to home and some of the things that happened to the Rudd family could, and did, happen to us on the farm when we were struggling.

After a diet of television and more high-class literature, fostered by the schools, I was doubtful whether my own offspring would appreciate the Selection tales, especially as living conditions have changed so dramatically. They have not known the battle of the farm, they have always had shoes to wear and they have never been short of tucker. But they have a strong sense of history, national and family, and they are intensely interested in anything that their early settler forebears may have experienced. So when a new edition of the Selection books appeared I bought a copy and casually mentioned that it was available. Modern adventures, the classics and other recommended reading were forgotten until the stories of Rudds were completed with much chuckling, loud laughter and extensive quotes to the rest of the family.

It appears that the Australian spirit still lives on in some of the teenagers of today. But they still lack the feeling of comradeship that we enjoyed as we sat around the fire and Dad read aloud to us those long winter nights, continuing the instalments the next night and starting another book immediately one was finished. The time we devoted to the discussion of the book, and the arguments we had over it, did as much for our education as anything else in school or out. Unfortunately, radio, television and other distractions of modern life, to say nothing of loads of homework and study, have caused an abandonment of the family reading circle of other days and we are all the poorer for the lack.

The Land, 10 August 1961

Perils of nature were no bar to Santa Claus

Santa Claus always managed to visit our farm irrespective of assorted crop failures, floods, bushfires, pests and other hardships at various times. The old gentleman got through everything, even a bushfire on one memorable occasion, much to the delight of the younger members of the family.

Christmas was always a big thing with us and preparations would start well in advance. So well in advance that the home garden would be planted with Christmas in mind. We must have beans and cabbage for Christmas and there must be a plot of new potatoes from which some could be dug for the dinner of the year. Sufficient fowls to last over the Christmas and New Year period would have been selected well in advance and brought to just the right stage of festive condition. Hanging in the dairy would be choice ham selected from the bacon we had cured in the autumn or winter. Hanging near it would be the pudding that had been mixed and cooked some time before, just ready to pop into the boiler for the final stage of cooking.

On appearance, those hams in the raw state were most unattractive, but they were delicious to eat after Mum and the girls had done their job with them and Dad had carved them into slices that were not too thin. The pudding, too, as it hung in its cloth, was not inviting in appearance, but it, too, ate well. In it would be an assortment of jewellery and small change, chief amongst which was a wedding ring. It was a tradition that the bachelor or spinster who scored the ring would be the first at the meal to marry. As we generally had some of our bachelor uncles as guests there would be plenty of fun as the pudding was eaten. A button was included to denote permanent single blessedness. The girls never seemed to want to receive that token with their pudding—they were more eager to get the ring. A golden horseshoe for good luck was always welcome but it was not as eagerly received as the silver coins by the youngsters.

But to return to Santa Claus. One of my sisters regarded Christmas as the most important part of her life and she still does after all these years. Even now, it is not uncommon in the early hours of Christmas morning to find her wandering around the house on an inspection of presents. When we were young we used to hang up our stockings. These made it easier for her to find out what she had got herself but when we graduated to a Christmas tree the job was a lot more difficult and she would generally fall over some of the furniture, waking all but the soundest sleepers.

There were always routine jobs to be done on Christmas morning such as milking cows and feeding pigs and calves, making sure the water supply for the animals was adequate, and so on. Led by Dad, we would be on the job early and clean it up in record time leaving the remainder of the day free for talking, eating and generally enjoying ourselves. Only essential work was done in our district on Christmas Day, except for one crusty old bachelor who always insisted on doing a couple of lands of ploughing. Even invitations to share Christmas dinner with various families were turned down by this latter day Scrooge who apparently regarded Christmas as just another day.

One thing regretted by most of the womenfolk was that no Christmas church services were held in our district. It was pocketed away from more closely settled areas and was probably too distant for any of the clergy to bother about a handful of settlers.

In one way or another we all managed to enjoy Christmas, to say nothing of the sports or race meeting organised for Boxing Day.

The Land, 21 December 1961

We all went to the bush dance

Dances are still held in country districts but they have lost much of the colour they had a few decades ago.

There would be dances run by tennis clubs, by various churches, for 21st birthdays and other milestones. In fact, little excuse was needed to organise a dance. This meant that with the balls staged by certain organisations and groups, there was something to do most of the time. Scene of these functions would be the local hall, a big hayshed or a woolshed. The latter was preferred because of the better floor. People would begin to arrive at the hall early, the women jealously guarding baskets of food in readiness for the supper that would be served later in the evening.

We would travel by horseback, horse and sulky and on foot. As the age of speed began to hit us, some daredevils would roar up on motor bikes. In due course cars and lorries would mix with the horse-drawn transport. But the really exciting times were when everyone used horses. A young fellow and his girl would ride up on their respective hacks, she more often than not being side-saddle, or a big family party would leap from an over-laden buggy drawn by a pair of horses.

Music was provided by locals, sometimes only a fiddler, but often accompanied by a pianist. There would be mouth organ soloists, sometimes a mouth organ band, but only for the most "toney" balls would a dance band with drums be engaged. That would be something for wide-eyed wonder by most of the young folk. Despite the sophistication the dance band added to a ball, I think we were happier when our music came from a fiddle and a concertina or accordion. I have seen men play the fiddle and the accordion for the whole evening with barely a break. The perspiration would be pouring off them but they would not give in; they would join in some of the reels as they played.

In those days the dances were waltzes, schottische, lancers, reels, barn dances, polkas and other old-timers. Not until later did the tango and two-step make their way into the bush.

Supper was one of the highlights of the evening and trestle tables would groan under the contributions that everyone present had made. As baby sitters were unheard of, every member of the family would be at the dance. This attendance of children and teenagers meant that no food was taken home. Adults ate well and the youngsters went through it like a plague of grasshoppers through a crop of green wheat, stripping everything as they went.

Because the whole family went along, a special room would be set aside for the babies and the mothers would take it in turn to keep an eye on them as the others danced.

The most important person at those dances was the Master of Ceremonies, or MC as he was better known. He would call the dances and make any necessary announcements. Some chaps made a profession of being MC and there would be great rivalry between them to get the job. At one dance in my early youth, the MC got up to make a special announcement. "There is a gold brooch lost in this hall tonight," he said. From a character at the back of the hall came a supplementary remark in a booming voice: "There is also a b—— hat." Fortunately, both missing articles were found.

Yes, the old time bush dance had a special part in our social lives. And many a romance blossomed as lads and lasses rode home together in the moonlight of the early morn.

The Land, 24 August 1961

When wagon was main transport

In the days before the carrying industry was mechanised, most farmers had their own form of transport for the products of the farm, generally in the form of horse wagons.

Wagons were of various types—the conventional, with boxed-in sides, front and tailboards, tabletops—and of various sizes, generally limited to the available horsepower. Loading a tabletop was a bit more difficult than the other type and the load had to be secure or it would move and probably fall off.

In our district, the main products transported some 12 miles to the railhead were potatoes, chaff and oats, which kept us busy in their respective seasons. Of course, the wagons were used for many other jobs, including carting-in hay, distribution of feed during bad seasons, carting wood, removal of stones and other rubbish from new cultivation land, carting an odd load of wool.

Our wagon, being a tabletop, had other uses when not mobile. It made a good work bench for rough carpentry, for instance, and had been used as a stand for the piper at a picnic when Highland dancing was a feature.

A necessary job with a wagon was to keep the wheels and the turntable greased. It was vital that the vehicle be kept in a shed or under some other cover in the hot weather or the timber on the wheels would shrink and the iron tyres would come off. When this happened, it meant a lot of work in the way of applying wet bags to swell the timber again, hammering wedges in between the felloes and the tyre, plus quite a lot of mental anxiety. Usually the shrinking process was not entirely satisfactory and a trip had to be made to the blacksmith to have the tyres shrunk on again. This involved cutting a piece out of the tyre, welding it again and then hammering it on while it was still hot. Being a tight fit, the hot metal tyre burned its way on to the felloes.

With the wagon in good order, it was loaded the night before a journey was due. The horses were harnessed before the crack of dawn.

On arrival at the railway station, groups of farmers would help each other to unload the wagons into the trucks. That job completed, a certain amount of shopping would be done after which the teams would head for home in batches of two or three. We always knew when ours was getting close because the old dog would prick his ears and show some signs of excitement although no one else could hear anything. But it was not long before we could hear the ironclad hooves striking the macadam as the team pulled the wagon up a steep pinch a couple of miles from home. That noise carried well into the frosty air of winter, the season when potatoes were carted. After the wagon had been backed into its shelter, the team would be unharnessed, fed and rubbed down. Then it was time for the family's tea when Dad would tell us about the day's trip and any unusual incidents that had befallen him and the team.

The Land, 12 October 1961

Ensuring the success of a clearing sale

A clearing sale is a big event in any country district. People buy a lot of stuff at such sales that they would never even look at otherwise.

A lot or organisational work goes into these fixtures. You sell your place, perhaps, or you want to get rid of a lot of the antiquated plant, old furniture, sheets of iron, used fencing materials, culled stock and other odds and ends, so you decide to hold a sale. A yarn with your stock and station agent will put you on the right track if you haven't had any previous experience.

The best plan is to get the various articles into some sort of order, even if it is only in heaps. Arrange the offerings in rows, with a fairly wide lane between each row to give the potential buyers and spectators enough room to move about and make their inspections, usually to the accompaniment of some sarcastic comments. If you hear some of those remarks, you'll be glad you are leaving the district even if you know that your neighbours have even worse collections of junk on their own places.

A lot of spectators come to a clearing sale, or they regard themselves as spectators when they come. Once there, however, many of them lose their spectator status because they just can't resist making bids. The success of a sale is mainly due to those people. They seem to get a bit feverish and make bids on stuff to send the prices up for people who really want to buy it.

It is a good idea to start preparing for the sale several days ahead of the advertised date. No matter how tidy you think you are, you'll muster up a lot of junk you didn't know you had, and most of it has a market.

If you are to offer stock, make sure they look reasonably attractive and that they are drafted into lots according to age and/or condition. Check the stockyard fences for stability because a lot of people will be climbing up them to sit on the top rail when the sale is in progress. There is nothing worse than for a fence to collapse throwing a rail full of people into the dust. That sometimes kills buying instinct.

Any harness for sale should be dusted over at least and oiled if you want to make it really attractive. Knock the cobwebs and dry grass off other offerings because the aim should be to make everything look as good as possible. But don't splash too much fresh paint around as newly painted articles arouse a suspicion that you might be trying to hide something. Everyone knows that we don't bother too much about painting things we use around the farm. If household goods are to be sold in the house, the women will look after them.

So, the preparations for a clearing sale generally resemble a dozen spring cleanings rolled into one.

In my younger days we reckoned it was a good thing to put on a feed for the buyers and those who just came along to "see how things go." Generally, the menu comprised lashings of corned beef sandwiches, some with mustard, some without. Cakes and scones were often added. Kerosene tins of black tea were necessary with every mug, cup or pannikin that could be pressed into service. Some hospitable people would turn on a "nine" or "eighteen", but that is not really a good plan because a lot of likely bidders would cluster around the kegs until the beer ran out, to the neglect of the sale and your ultimate profits. Besides, there is always a danger of fights developing if beer is available. But it is a good idea to have a bottle of, say, whisky in some private spot with which to refresh the auctioneer, the bank manager and other important and influential people who are bound to be there.

A lot depends on the auctioneer if a clearing sale is to be a success. If he is the jovial type and throws himself into his work, with a lot of witticisms, he'll have the crowd with him and the bids will come thick and fast. He gets things started and mob psychology does the rest. People get carried away and lots of them will leave as the new owners of articles not as good as similar junk they have at home. They'll realise this when their nerves quieten down, but meantime you're sold out. Some people, for instance, just can't resist the appeal of a small roll of battered wire netting, an old scuffler, a bridle without a bit, or tools that have seen better days. But they all have a good time at your clearing sale.

Of course, if you attend other people's sales, the odds are that you'll be a buyer too, even if you only go along to have a yarn with people you know and to get an idea of the values. It cuts both ways.

The Land, 23 March 1961

No summer uniforms for mounted police

We read in the papers recently where a magistrate had taken exception to a policeman appearing in court dressed in the summer uniform of shirt and trousers with, of course, tie and shoes. Although it was a hot summer day, the "beak" wanted everyone in the court to wear a coat. That, of course, was his business, but it reminded me of the way policemen dressed when I was boy.

In the bush we saw mainly Mounted Police, except on our rare trips to town when the policemen on foot appeared tame by comparison.

Mounted policemen paid us irregular visits to collect stock returns, inquire about cattle duffing (a major offence), or just called in for a cup of tea and a yarn. They would have suited our magistrate friend, for they wore short, tight jackets buttoned to the stiff celluloid collars that showed a fraction of an inch above the dark blue of the jacket. And they wore white riding breeches with highly polished boots and leggings; spurs, of course. Their mounts were among the best in the district. They were well groomed and the saddlery shone so that you could almost see the reflection of your face in it.

Those white trousers intrigued us. They were smart—until the policeman dismounted on a hot day. Then, and especially if he was on the plump side, the seat of the trousers would be stained with perspiration where they had been in contact with the saddle. This, of course, distracted from the smart appearance of the wearer. So much, indeed, that it tickled our fancy and it was difficult to hide the grins. Hide them we did because we were scared of the majesty of the law as represented by a young mounted policeman who probably wished he was nearer the fleshpots.

I'll never forget the visit of a policeman late one evening. He was on a mission that notched up another dislike for the uniform where I was concerned. He called to tell us that the local school was re-opening after some years because there were now enough children in the district to warrant the Government getting it going again. Accustomed to the sketchy lessons from our parents, we were not looking forward to regular school attendance, which we likened to breaking-in a young colt. Where this eight-year-old was concerned, the introduction of the new way of life could be blamed on the man in the white riding breeches, so I immediately abandoned any sneaking ambition I may have had to join the Mounted Police.

The Land, 2 March 1961

Primitive conditions in some old-time dairies

Conditions on dairy farms were crude before various authorities became interested in the industry. Their interest stemmed from a desire to improve quality and to ensure that milk, cream and butter were produced under conditions that would not endanger human health.

Some of the early farmers did not seem to understand such high motives, as their dairy premises indicated. Cows were milked in sheds remarkable mainly for the fact that they had three walls, which protected the milkers from the wind and a roof which kept the rain and sun out.

Earthen floors had a habit of eroding away and it was not long before the cow was standing in a depression. Some venturesome souls would concrete the floors of the sheds to ensure that the cows remained on the surface of the soil and did not vanish entirely when they were standing in the milking position. Of course, the manure and urine made things unpleasant for all concerned and this possibly had a great deal to do with the swing to concrete floors. Bails were made from split timber put together in true bush carpenter style, and the big yard was of similar construction. In those days no one thought of holding yards—the cows were run into the main yard from where they were drawn as their turn to be milked came, and they went back to the yard after milking.

Usually the walls of the cowshed were of slabs with generous spaces between each one, through which the winter winds howled. They were cool enough in summer. But at that season there were other disadvantages in the old time dairies. These included a high odour to which the milkers became accustomed, I suppose, but they were definitely "on the nose" to anyone of finer susceptibilities.

In rainy times the cow yard and shed became nightmares as the animals churned up the evil smelling mud which they carried into the bails with them. As there was no regular water supply to which a hose could be attached, it was an uphill battle with broom and shovel to attempt to keep the bails clear of mud at such times.

And when milking was finished, the cursory wash out with a couple of buckets of water, given by the more fussy farmers, made little impression on the collection of mud. Where no water was used, the bails were really something and their only clean periods were when they dried out and the muck could be shovelled and swept away. Generally, it just went into the yard. It was transported to and from the bails again in due course, with a lot more new matter on each trip made by a cow.

No, those old dairies were far from hygienic and by present standards must have been breeding places for millions of germs. How the population that consumed the dairy products survived is a puzzle, but the survivors may have become immune. It is a fact that contagious, milk-borne diseases were much more common in those days—a strong argument in favour of the strict supervision of the dairying industry that was eventually introduced. Unfortunately it was many years before many dairy farmers followed the move towards hygienic conditions. This was one reason why the quality of our dairy products took so long to reach a high standard and why their keeping ability was not good.

Where dairy products are produced for sale, premises are now far removed from the old-time germ incubators. But on some places where a few cows are milked for home use you'll still find the same set up.

The Land, 16 November 1961

Milkmaids have always made excellent wives

In the days before the almost general adoption of milking machines, a steady hand milker was worth a lot to the dairy farmer.

Generally, girls made the best milkers and a dairyman with a big family of daughters was well situated. Another angle was that when a young man went courting, he generally selected a girl who was a good milker and it was not always necessary for other things to be equal.

Glamour was absent from many of the farm girls, but what they lacked in that regard they made up for in their ability as cooks and housewives, occupations for which they were trained from early childhood. In fact, they learned how to care for a home—and a husband—as soon as they learned to milk, and that was often before they started school. We used to hear a lot about "child slavery" because of children working on dairy farms, but milking cows and feeding calves could not be described as hard work.

Possibly, the greatest disadvantage was the necessity for the family to rise early, which had some effect on their schooling, as they would get sleepy before the day was over. Generations of dairy farm families lack nothing in health and intelligence. Although the ability of girls to milk cows may not be considered so much by potential husbands these days, the wives are good housekeepers and they know as much about running a farm and putting the profits to good use as their husbands do.

It all depends, of course, on what you are accustomed to. The environment of a dairy farm, in a district where little else than dairying was the main occupation, suited the people who lived in it. The only snag was the need to be on the job seven days a week, which became a bit wearing. But there were compensations. Sport and social occasions could be sandwiched in between milkings and in the evenings, which helped to give a full life to folk from the dairies.

An attraction of dairy farming has always been the regularity of income. That monthly cheque from the factory has been appreciated by thousands of families since the earliest days and it is an advantage that few other rural families have enjoyed. Regular payments to farmers have helped to put stability and solidity into the towns in dairying districts where business people have not had to wait for payment longer than the monthly account. It may have been different in the pioneering days but in my time it deserved and received serious consideration.

My mother, born and reared on a dairy farm, always advocated getting a few cows together when things were tough. Although we were in a mixed farming district, we found that dairying fitted in well with the other branches of our activities. It is certain that on several occasions the cows saved us, and others, from insolvency. People may not want potatoes, the bottom might fall out of the bran and hay markets, but people always need dairy products and the stability of the market gives the farmer an opportunity to budget for times ahead.

Girls reared where such conditions rule get a sound grip on the economics of the situation. That is why they still make good wives, even though they may have left the farm to become teachers, secretaries, models or to follow some of the other glamorous professions open to them today. Some of the meals those dairy wives prepare for special occasions would set back even a top-line gourmet. It is always a pleasure to receive an invitation to lunch with a dairy farming family at a show—far better than most official luncheons.

Did I marry a girl from a dairy farm? No, but I have no regrets because my wife came from an old English farming family.

The Land, 23 November 1961

When big change came to the pigs

Pig raising has made many advances, but on lots of farms you'll still find the animals confined to small sties where conditions are unpleasant and, one would think, uncomfortable, especially in wet weather. Food is still thrown over the fence to the pigs and a dish in the corner is filled with a ration of milk once or twice a day. Household scraps, usually in the raw state, form the bulk of the feed that pigs kept under these conditions receive. That was the way in the old days, when pig keeping was mainly to augment the family diet with pork and bacon.

Apparently the method was a heritage from the old Irish days and ways, brought to Australia by the early settlers. Of course, one cannot blame the Irish entirely for the guide at Captain Cook's Cottage in Melbourne will tell you that the Cook family raised pigs. And the largest room under the cottage roof is the one in which the pigs lived. A wall divided them from the family quarters, one is pleased to note.

When the revolutionary idea was first advanced that pigs could be raised with success, and at comparatively low cost, on pastures with a wider range, it was met with scepticism by most of the old time pig men. Their view was that then pigs would run condition off on the wider range and that they would not fatten quickly. Some even advanced the contention that pigs needed dirty conditions to keep them healthy. Any idea of giving good pasture over to pigs and of sowing crops, such as turnips and mangolds especially for them, was treated with disdain.

It often astonishes me that the new system made any progress at all, in view of the welcome it was given. But I suppose a few foolhardy souls in a district gave it a kick-off, did well, and were followed, rather reluctantly, by their neighbours over a long period.

In our district we were among the first to get pigs out in the open. We fenced of a paddock for them, subdivided it and put a portion under improved pasture, another under turnips and left the third in its natural state. Our pigs thrived once they got used to the new way of living. We found that they actually improved the soil for crops. Feeding methods were revolutionised and we added a fair proportion of grain (which could not otherwise be sold for profit) to the ration, boiling it before it was fed.

One of the main ingredients of the ration was a liberal allowance of separated milk. Dad warned us about the milk. We had to be sure that the pigs did not gulp it and it was not to be fed until at least an hour after separating. The idea was to allow the aeration to subside because a lot of newly separated milk would not be the best for the pigs.

We observed the instructions over a short period, but there always comes a time when a bit of a rush will cause you to ignore the rules and "give it a go" to get the job done. On this particular evening the job of feeding the pigs was mine and I aimed to hurry it through. I sloshed the separated milk into the troughs and the younger pigs sloshed it into themselves almost as fast as I could pour it. Then I forgot all about the pigs until Dad arrived home an hour or so later. There was the devil to pay. He stood at the fence of the pig paddock and roared. When we rushed out we saw those white spots glistening on the pasture under the moonlight. Each of the spots was a young pig, burst because of an over supply of separated milk. That, of course, cut the profit on that year's batch of pigs.

The Land, 14 September 1961

Coach painter was a master craftsman

Another craft that has vanished from the country districts with the march of progress is that of the coach painter. No coaches run today, and there are no horse wagons.

Modern transport is mass produced and the application of paint and/or enamel is a factory job, applied with modern sprays and similar equipment. But less than half a century ago, sulkies, buggies, drays, carts and wagons, were painted by hand, usually with skill and with materials that lasted for years. It needed a steady hand to do the "lining"—those thin lines of decorative paint on various parts of the lighter vehicles and even on some of the wagons. Lining was in a contrasting colour to the paint used for the bodywork and it lifted the solid colours as contrasting accessories do with high fashion clothes.

Some of the more glamorous sulkies and buggies would be stained to a "natural" tone, or the raw timber would be varnished to bring out the attractive graining and colouring apparent in the natural state. It was then that the "lining" would provide an attractive contrast, and great pains would be taken to ensure that the colours chosen would tone in well, or contrast sharply with the body colour. Black was the other fashionable body paint. With it, the painter had plenty of scope for relief with his chosen "lining" colours. A bolder approach was used for drays, spring carts and wagons. Body colours would at times be somewhat startling and the relief would be provided with even bolder tones used with some freedom.

Artistic designs were popular on the heavier vehicles. If the painter fancied himself as an artist they would be complex. Some painters even went to the extent, at the suggestion of the owner, of painting scenes on various panels. In painting the name and address of the owner, the range was from the genteel lettering of visiting cards to letters of a size that, as Dad used to say, could be seen by a blind man with a cork eye on a galloping horse.

One of the aims was to have the wagon painted as differently as possible to the other similar vehicles in the same district. This was easier for an ingenious painter who handled the work for all the farmers of the neighbourhood but was sometimes difficult when a wagon was bought from outside. Then, if the colour scheme and design resembled too closely that of another wagon in the district, a re-paint job would be ordered—and hang the expense.

Not all vehicles, however, were kept up to the mark. You'd find some with the paint peeling off and some that looked as if they had never felt the touch of a brush. But if someone turned up with a new, or freshly painted, wagon on trucking day at the station, you could bet that it would not be long before there was a wave of regeneration.

Sometimes the owner would do the job himself, but the do-it-yourself efforts were rarely satisfactory. Painting was a job for a craftsman, usually a man trained in signwriting or "lining" and who took a professional pride in his work. And, like the true artist, he'd always sign his work—neat lettering giving his name on some inconspicuous part of the vehicle.

The Land, 3 August 1961

When politics was a way of life

There are all sorts of ways in which you can get to know how the other half lives. One of the best is in the maelstrom of politics, especially in an election campaign. This was never more apparent than during the election battles of the Depression years when politics really meant something to everyone and a fight would start at the drop of a hat.

Take Bill Lew, one of the battling stock and station agents in our town who, with some of his cronies, threw his not inconsiderable weight behind the Country [now National] Party man. Bill's forte was the holding campaign—he'd open a meeting and talk until the candidate arrived anything from 10 minutes to an hour or so after Bill started.

He'd cover every subject under the sun, with special reference to "Jackie" Lang [J.T. Lang, Labor Premier of NSW from 1925–27]. "[Jackie] has got his point of view, but it isn't ours. We're all on the fringe of the underworld, just a couple of spuds removed from the dole, and Jackie will keep us there as long as he can," Bill would tell his audience. Then the heckling would start, but it never bothered Bill—he'd sail majestically on, shouting at the top of his voice (easily the loudest in the town) until the puny opposition gave up.

But the thunderous sentences and the fighting words he enunciated would draw the crowd, many of whom came along for the show when they knew that Bill was a speaker. And he always spoke out-of-doors because the other two parties had got in early and booked all the halls in town. As Bill said, however, you couldn't have much fun in a hall and when the fighting started something was bound to get broken and someone would have to pay for it. Whereas, at open-air meetings, anything could happen and mostly did.

If the fights got too willing, someone would tear a few palings or pickets from a fence for use as weapons, but the majority frowned upon that sort of fighting and joined forces to overpower and disarm the really violent. Fortunately, the New Guard [A militant conservative organisation opposed to what it believed was the left-wing drift of the community during the 1930s—Ed.] had not extended its organisation to our area, but there were other hazards and one night supporters of both our opposing candidates joined up to wreck our meeting. They yelled and they coo-eed, but Bill kept on in masterful style, his booming voice echoing over the lesser noises. Faced with the fact that they could not stop him short of an attack on his person, the disruptionists started several brawls in the crowd. The police arrived but the meeting had been wrecked from a political point of view.

Bill decided it was time to teach the others a lesson so he called his closest associates together and planned a campaign. As it happened, one of the opposing candidates had a meeting in the town's biggest hall the following night—and the hall was packed, with Bill's mob placed strategically in the crowd.

All went well until the candidate himself got up to speak. He had barely got into his stride when Bill rose majestically and, for effect, mounted his seat. "Mr Candidate," he said, "before you go any further, can you tell us whether you believe in the British Empire?" There was stunned silence for a moment, and then, as they say in the classics, pandemonium broke loose.

In more sober moments the question would have brought a quiet affirmation and possibly a laugh from the crowd. But feelings in those days were tense and about a dozen people leapt to their feet and expressed their opinions of Bill's manners in interrupting the speaker. At the first lull, Bill's voice boomed through the hall: "The question is still unanswered."

"Sit down you big mug," someone yelled.

That was the signal for Bill's own mob to go into action. In all sections of the hall they rose and yelled catchcries such as "Fair Go!" "Free Speech!" "This is a democracy and we want an answer!" "Is your man a red-ragger?" and similar queries.

"Do you believe in the British Empire?" Bill kept booming.

The poor devil of a candidate could not make himself heard and although we knew that he was as loyal as any of us, it suited us better to keep the turmoil going than to give him a chance to be heard.

Naturally, someone threw a few punches and there was a rush to remove the fighters, willing civilians beating the few police by yards to the centre of activities. The fighting section, pressed by new arrivals, moved out from the hall and through the vestibule into the street. Our crowd left the clotted humanity and re-entered the hall where Bill and a few others were now demanding the essence of British justice to have a simple question answered. We kept the uproar going until a burly police sergeant pushed his way to Bill and asked him to stop.

"At the request of the police, I'll resume my seat," roared Bill, sitting down, "but I still want to know if he believes in the British Empire."

Yells and roars exploded again, although the crowd had thinned appreciably: the excitement was now outside. The chairman was not a strong citizen. It was soon obvious that the meeting had got beyond his control, so it was not surprising when a couple of dozen fellows stood up and sang the National Anthem that everyone joined in—and the meeting was over.

Those were the days when politics was a way of life, when meetings were crowded, and when political speakers provided better entertainment than the picture show.

The Land Annual, 18 October, 1961

Poor start checked sheepowning career

Did I ever tell you how I first became a sheepowner? It happened in my preschool days due to the generosity of my maternal grandfather. But it did not represent the foundation of a fortune for me, nor did it qualify me to become a member of either the Graziers' or the Sheepowners' Associations.

My grandfather presented my two sisters and me with a lamb apiece. It was prior to the formation of the Junior Farmers' movement, but we cared for those lambs in a style that would have won us prizes for projects if we had been doing it a couple of decades later. And the lambs thrived.

Came the first shearing, an eventful period for the whole family. My father's flock was shorn with the old hand "tongs" and our pets were given the same treatment with a little more care than the older sheep received. After the wool had been baled and sent for sale, we waited anxiously for the returns. When we got our share of the cheque, three new savings accounts were opened.

In due course the lambing season arrived. The pet sheep belonging to my sisters each produced twin lambs, which brought an appreciable increase in their flock numbers. My sheep was a wether. Thus my flock did not grow.

For revenue additional to receipts from normal sources—and they were not numerous—I had to depend on my annual wool clip of one fleece. But the flocks owned by my sisters continued to increase each year. Those pet ewes produced more twins than single lambs and their progeny did the same. This meant that the girls had bigger wool clips each year plus profits from the sale of wether lambs. Their bank accounts grew and they lived in comparative luxury without having to follow the hard life that was mine—rabbiting, picking up "dead" wool and bones in the paddocks, helping with potato planting and digging and other odd jobs. Even then my income was lower than theirs. This went on for some years until, for various reasons, my sisters sold their flocks. My lone sheep died of old age.

In the years since then I have become convinced that I lack business sense. If I had possessed any, the natural thing would have been to invest my annual fleece returns in ewes. That, with a bit of luck, could have made me a big flockowner in time, with more than a passing interest in reserve prices, promotion and synthetics.

The Land, 27 April 1961

Tracking bees to their home

My Dad had a marvellous eye for tracking bees to their "nests." After a day in the bush with him, it seemed that he could see a bee for miles.

This bee-tracking was an annual event to which Dad, in particular, looked forward with a great deal of pleasure. The idea was to watch bees watering at the edges of a dam or stream and, when they had taken their fill of water, to keep an eye on them as they flew towards their home tree or other honey storage spot. Naturally, the bee's navigation was not really good when the insect was weighed down with water. That made the job of tracking them a little easier.

It was necessary to mark the route taken by the bees with trees, rocks or other formations then set off in the same direction, either on horseback or on foot. You might travel upwards of a mile or two, but if the direction was correct, and it nearly always was with Dad, you'd come to the tree where you'd see the bees flying in and out of a hole in the trunk or a limb. The location of that tree would be marked in your mind and you'd be off to track some other bees. This might occupy the best part of the day, taking out time to boil the billy two or three times.

Towards dusk the next phase of operations would begin. We would adjourn to one of the trees the bees were using, light a smoking fire and cut into the trunk or limb near the entrance hole. There would be a lot of tapping and listening to ascertain where the bees were working. Sometimes it would be a major job. The tree might be a hard one to cut, several cuts might have to be made to trace the honey, or the wood might not split. But it was only on rare occasions that the tree was abandoned without sighting the domicile of the bees. The object of the smokey fire was to keep the bees quiet and as protection we would be muffled up with large handkerchiefs or veils. Even then someone would be stung when he least expected it.

When the opening had been completed, the honeycomb would be removed, generally with bees, and placed in one of the kerosene tins we had brought along. Our work would continue until we had gathered what we considered enough honey to last for about a year. This yield would have to be divided as one or two neighbours would usually be in the party. By the time the last of the honey had been gathered, all members of the party would be red-eyed from the smoke, somewhat sticky from the honey (something I hated), sore from stings, tired and irritable.

At home the next day, the honeycomb would be placed in a series of cloths, hung from a rafter in the shed and allowed to drip into cans below. This separated the actual honey from the comb, the odd bees we had gathered and a multitude of splinters and other foreign matter from the trees.

One of my worst efforts was when, on the way home, I was chewing the honey from a piece of comb. Somehow or other there was a loose sting in that mouthful and my tongue swelled considerably, preventing me from talking or eating and making even drinking difficult for a couple of days. Ever feel that you had tennis balls in your mouth? That was the sensation.

The Land, 27 July 1961

More are reaching the century these days

It is becoming common to hear of people living to 100 years these days when those in their eighties and nineties seem to be in the youthful bracket. This is because our life expectancy is growing and must now well exceed the Biblical level of three score years and ten. Yet I read that in some of our neighbouring Asian countries it is about 40 years. However, more people can look forward, bar accidents, to seeing the century out, or going close to it, than 20 years ago. Much of this must be put down to medical science, which has learned a lot about coping with killer diseases and other complaints that shortened the life of man. Paradoxically, war, one of the greatest killers of all, always yields knowledge that helps in the fight against disease and death.

When I was a young fellow, anyone who lived to be 100 was big news and was generally a coddled invalid. Nowadays, however, the picture is different. On television the other night I saw pictures of a woman celebrating her 101st birthday and she was as active as many women 20 years her junior. You could see that she would not appreciate coddling. Then there was the case of a man who had reached 100 and, apart from a walking stick, he could go for a stroll or weed a garden with as much aplomb as you or I.

When I was going to school, an old Irishman neighbour, who lived about three miles from us, attained his 100th birthday. He had been bedridden for years but that had not affected his mental faculties. He had a good grip on current affairs, particularly on the "troubles" in his native Ireland. He attributed his long life to plain food, copious draughts of the best overproof rum (he reeked of the stuff) and pipe smoking. In his opinion, anyone who smoked cigarettes was committing suicide at an early age. That was before any publicity had been given to lung cancer and associated troubles, but the old boy was definitely against the cigarette.

His opinion is shared by an uncle of mine, who is nudging 85 and looks like going on for another 15 years. My uncle, who prefers beer to rum, says the pipe is definitely the smoke for health. To support his claim he says he has been smoking since he was 12, and always a pipe. "I've had an occasional cigar, but I've never smoked a cigarette," he told me the other day when advising me to concentrate on the pipe myself. "Pipe smoking never did anyone any harm." He said nothing at all about two members of my family who had died in their seventies—non-smokers both. But they probably had other vices, although they were not as wild as the senior uncle.

This and similar observations incline me to the view that longevity may be due to some inherent physical strength in the people concerned. You will see a couple of members of the same family live to an old age, but others will die much younger, something that has occurred in my mother's branch of the family.

A more enlightened medical science provides some of the answer, inherent physical strength some, careful living some more and, probably, good luck is the deciding factor. And, as someone told me several years ago, the best way to live to a ripe old age is to develop a chronic complaint and look after it.

The Land, 5 April 1962

Highlight that history missed

I notice that the Bank of New South Wales [now Westpac] celebrated its 145th birthday the other day, a milestone for Australia as well as for the bank.

Considerable publicity has been given to the fact that the bank's first depositor was one Sergeant Chisholm of the New South Wales Corps, whose descendants have thrived in this country and who have been friends and neighbours of my own family for more than a century. But no publicity has been given to the fact that my great-grandfather was the first man in Australia to be refused an overdraft by the Wales.

It was a grim day for the family, although the term may not have been "overdraft." That, I think, came into actual use at a later date. It was probably a loan of some kind for one of the harebrained wildcat schemes for which my great-grandfather was well-known throughout the colony and for which he had extracted various amounts from certain of the citizenry at appropriate moments.

One can visualise the stern directors of the bank considering the request for financial aid and discussing the possibility of repayment. Perhaps they did not consider it too long, because my great-grandfather's activities must have been known even in those rarefied circles. At any event, they knocked his application back. This, of course, precipitated a crisis. Instead of our family being numbered among the early supporters of the Wales, our names were missing from its records for several years.

Somehow or other, my colourful ancestor made arrangements for finance from some of the early merchant princes of the colony. He did reasonably well, apart from his farming activities along the Hawkesbury. Having built up his finances, repaid some of the money he owed to various people and devoted his time to community service, he was looked upon as a sterling citizen, a point that the directors of the Wales were not slow to appreciate.

When the next request came from my great-grandfather, the directors of the bank had no hesitation in granting most of what he asked for, and they had no reason to regret it. Unfortunately for both parties, a further deal was hit hard by floods on the Hawkesbury and drought conditions elsewhere. It was years before the family could afford to have any dealings with any bank.

It has gone like that ever since. The fortunes have been up and down, the variations depending mainly on the human element but being tempered by such natural conditions as floods, droughts, bushfires and other disasters. However, one bank manager has been known to say that we would be a wealthy family if we did not get so many wildcat ideas and if we were content to take a little guidance from our bankers before the position became really serious. He might have something there, but as great-grandfather would have said, we'd have missed a lot of fun and would never have been the first family to be refused an overdraft by the country's first bank.

The Land, 10 April 1962

Train trip that lived in memory

As you look back on your past life, there is one journey that stands out as the most memorable of all. It may have been long ago, or it may have been recent, but there will be something about it that sticks in your memory, thus making it an important part of your life.

During my career, I have travelled many thousands of miles by various forms of transport, ranging from horseback to modern airliners, but the journey I remember most is by train, the operative control being the NSW Government Railways. It was a trip undertaken prior to my fifth birthday. The family moved from my birthplace out west in the Cobar district to the South, near Goulburn. Here the wanderings of my branch of the family were to cease for some years after more than a century of nomadic adventures mainly behind stock and/or in search of that elusive metal, gold.

Involving the transfer of a family from one section of the State to another, that journey was, of course, a major operation and one that was bound to make its mark on the impressionable mind of a young child. Days were spent in packing up, but the time arrived when we boarded the train, a great adventure, but one that was to pall before the trip was over.

In those days, trains, especially the second class carriages, were built for utility rather than comfort and you sat up in a narrow compartment with your back straight against the wall with little room in which to move. Such cramped quarters had little appeal for active youngsters, and we were soon tired of it.

Early in the journey, the sun blazed down and we felt as if we were in an oven. As the carriages had no fans, all we could do was get rid of unnecessary clothing, open the windows and drink water, the supply of which had to be renewed at almost every station.

There was not much in the scenery to interest us, and the day was too hot for food to be enjoyable, but somehow or other the time passed. My recollections include the excitement of watching the locomotive taking on water, an occasional inspection of tickets and an inspection of the people who left or entered the train at the various stations. One thing I do remember about those people was that the women wore skirts, generally black that swept the ground, hats of great width to protect them from the sun and, for additional protection, most of them carried wide, black umbrellas, or parasols as some people called them. Some of the more finicky women would gather up their skirts with one hand, lifting them a couple of inches above the dust, as they held their umbrella aloft with the other hand.

As night fell, changes came, particularly in the temperature, which was little short of freezing, or that is how it felt to people who had come from the hot West. Travelling in a darkened carriage quietened us down for a time and we eagerly looked out as the train rolled into a lighted station.

Some time during the night we reached Blayney, the junction where we changed trains from the Western to the Southern lines. My first memory of Blayney is of cold, hard waiting room seats and bitterly cold weather conditions. My sisters seemed able to sleep but I'd doze off for a few minutes and then be awakened by the cold.

Looking back on it all, I often wonder that I did not start my usual practice of bawling when things were not to my liking, but I remained silent apart from a few grumbles and grizzles. Time passed, as it always does, and our train for the South came in. It was like a haven for a tired young lad and I curled up and went to sleep on my section of the seat, not to awaken until we were greeted by my grandfather on the platform at Goulburn.

In the years that have passed, I have developed feelings of respect for my parents, particularly my mother, because of the difficult conditions of travelling so far by train with a young family. They must have been impelled at times to throw us out the window.

The Land, 2 August 1962

March of progress cuts historic links

As I came through the old settlement of Parramatta this morning, I thought of its place in the history of Australia, of the many buildings still standing that link us with the earliest days of the then colony. But the march of progress is reducing the number of those links, just as it is cutting them down in Sydney Town and in other parts of old Australia.

Fortunately, more and more people are becoming interested in the history of their country, as is shown by the demand for books about the pioneering days and the interest shown in television features, even if that interest has, at times, been somewhat hostile. It is good that Australians are showing such interest, but it is to be hoped that their interest will encourage them to take action for the preservation of the history of their own country.

Many families can now boast the sixth Australian-born generation, but they don't know a lot about the history of their family, their district or their country. This ignorance is pretty general, however, and I am one that must plead guilty. For years I have passed regularly a place called Werrington Park on the Western Highway, but it was only the other day I learned the property was once owned by Mrs Mary Putland, daughter of Governor Bligh. I daresay thousands of others who see the place regularly don't even know that.

Visiting such old cities as Goulburn and Bathurst you can see, in the older areas, houses that were built in the earliest years of white settlement. They are small and, to the modern eye, entirely unsuited to the conditions of the district concerned. But they were solidly constructed, with their thick stone walls and heavy, sawn timbers. Unfortunately, it won't be long before "progress" demands the demolition of those places to make way for modern cottages and flats. One can hope that citizens imbued with a sense of history will be able to persuade local government authorities to take steps to preserve at least a few of those old houses.

Go out into the country a bit and the same story is repeated. Once gracious old homes, long since abandoned, are falling into decay. That arouses some hostility in me because it will possibly be only a few years before the more wealthy of us will, as has happened in other countries, think it fashionable to buy the old home of a pioneering family and restore it as a residence, if only for weekends. So much for bricks, stone and mortar, roughly hewn timbers, wattle and daub.

There is another side to the picture—the written record.

All over Australia there must be letters, diaries, account books and other records tracing back to the earliest days and telling the histories of pioneer families and long-established stations. Many of these records, priceless to the historian, have been carelessly destroyed. Some have been handed over to a local historical society or the Mitchell Library for safe-keeping and for the use of students, but others are still in cupboards, boxes, old tubs or tins where they were flung years ago and more or less forgotten. Many appeals for such papers have been made by various organisations, with varying results, and I make no apology for adding my humble voice to this chorus, for all such records help us to piece together the history of our nation.

In recent months I have had the good fortune to read a couple of books on the pioneering days. One tells a graphic story of the deeds of a grazing family and its remarkable trip across the continent. The story is told by a member of that family whose sources were written records and the personal memories of the old hands. That is one section of history that has been preserved. The other book, one of much greater detail, tells the story of a district through the work of its settlers and it was compiled in many hours of research involving the study of family records in Australia and Great Britain. It, too, has preserved for us a valuable portion of our history.

If you belong to a pioneering family, you can probably lay your hands on records that are as valuable, possibly of even greater value.

The Land, 9 August 1962

EPILOGUE

When *The Farmer & Settler* announced in August 1943 that my father had been appointed to its Editor's chair, the paper noted that he had "a comprehensive knowledge of all phases of agriculture and stock raising". In reporting the appointment, the *South Coast Register* said he was well-known to the farming community and that his "facile pen" would be of even more benefit to the man on the land now that he was Editor.

Jim Mahoney held the Editor's chair at *The Farmer & Settler* for a decade; after leaving it, he did not return to agricultural journalism for almost another. But when he did, he wrote the "A Bushman Remembers" series for *The Land*. As "Tarboy", he used that knowledge, his experiences of growing up and working in Manning Clark's "mighty bush", and that facile pen to write about an era of Australian life that has now gone for ever.

The "Bushman" stories reflect a daily rural life not much changed from that period of colonial history trolled by Russell Ward to search for the origins of what he called "the Australian legend". Bushmen, wrote Ward, were "above all, 'practical' men". A bushman, he said, "was, almost axiomatically, a man who could turn his hand to most tasks". This collection deals with many of those practicalities of bush life and is my father's attempt to ensure that the lifestyle of our forefathers will not be lost for ever.

—JSM

James S. Mahoney, BA (ANU), FPRIA

After a career in journalism, in which he reported agriculture, courts, transport and national politics for the *Sydney Morning Herald*, James S. Mahoney directed public relations programs for the Australian Mining Industry Council, the Australian Electoral Commission, the Civil Aviation Authority and the Australian National University.

He has a small consultancy, writes, and is a public relations tutor at the University of Canberra. He has been a guest lecturer in public relations at the University of Canberra, Deakin University and the Centre for Continuing Education at the University of Sydney, and has been an invited speaker on public relations planning and evaluation at international and domestic conferences; he has published internationally on the same subject.

James S. Mahoney has been actively involved in the Public Relations Institute of Australia, serving terms as Chair of the ACT chapter's Education Committee, as a Councillor and as its Vice-President. In 2001 he was elected a Fellow of the Public Relations Institute of Australia.